In the Collected Studies Series

MICHAEL TALBOT
Venetian Music in the Age of Vivaldi

RICHARD SHERR
Music and Musicians in Renaissance Rome and Other Courts

DAVID CHARLTON
French Opera 1730–1830: Meaning and Media

ERNEST H. SANDERS
French and English Polyphony of the 13th and 14th Centuries: Style and Notation

ROGER BOWERS
English Church Polyphony: Singers and Sources from the 14th to the 17th Century

RUTH STEINER
Studies in Gregorian Chant

DAVID FALLOWS
Songs and Musicians in the Fifteenth Century

RICHARD L. CROCKER
Studies in Medieval Music Theory and the Early Sequence

VARIORUM COLLECTED STUDIES SERIES

Music, Science, Philosophy

For Michael

Jamie C. Kassler

Music, Science, Philosophy

Models in the Universe of Thought

Routledge
Taylor & Francis Group
LONDON AND NEW YORK

First published 2001 by Ashgate Publishing

Published 2019 by Routledge
2 Park Square, Milton Park, Abingdon, Oxon OX14 4RN
605 Third Avenue, New York, NY 10017

Routledge is an imprint of the Taylor & Francis Group, an informa business

British Library Cataloguing-in-Publication Data
Kassler, Jamie Croy
Music, Science, Philosophy: Models in the Universe of Thought.
(Variorum Collected Studies Series: CS713).
1. Music – Philosophy and Aesthetics.
I. Title.
780.1

US Library of Congress Cataloging-in-Publication Data
Kassler, Jamie Croy
Music, Science, Philosophy: Models in the Universe of Thought / Jamie C. Kassler.
p. cm. – (Variorum Collected Studies Series: CS713).
Includes bibliographical references and index.
1. Music – Philosophy and Aesthetics. 2. Music Theory – History. 3. Music and Science History. 4. Model Theory.
ML3800.K187 2001
781.1'7–dc21 2001022175

ISBN 13: 978-0-86078-862-1 (hbk)

VARIORUM COLLECTED STUDIES SERIES CS713

CONTENTS

PREFACE

No part of knowledge can stand wholly disconnected from other parts of the universe of thought.
W. S. Jevons, *The Principles of Science* (1874)

In 1966–67 Sir Isaiah Berlin set out the following task for philosophy:

> to extricate and bring to light the hidden categories and models in terms of which human beings think (that is, their use of words, images and other symbols), to bring to light what is obscure or contradictory in them, to discern the conflicts between them that prevent the construction of more adequate ways of organizing and describing and explaining experience (*for all description as well as explanation involves some model in terms of which the describing and explaining is done*); and then, at a still 'higher' level, to examine the nature of this activity itself (epistemology, philosophical logic, linguistic analysis, and to bring to light the concealed models that operate in this second order activity itself).[1]

These words, which I read first in 1979, provided the encouragement then needed to pursue an approach taken previously in my doctoral dissertation, where I was concerned, in part, to extricate the models in terms of which some music theorists described and explained music-theoretic reality.[2] After 1979 I also began to investigate how some theorists in other, non-musical domains utilised models from the domain of music to describe their respective realities. A full-length

---

[1] I. Berlin, 'The Purpose of Philosophy', *Cogito: Journal of the University of New South Wales Socratic Society* (1966–67), 1: 3–12, italics mine.

[2] J. C. Kassler, *British Writings on Music, 1760–1830*...(Columbia University doctoral dissertation, 1971; Ann Arbor: University Microfilms #72–1337).

study, resulting from this investigation, was published in 1995,[3] although most of the following chapters also exemplify some of the ways in which I have interpreted the first part of Berlin's task.

It seemed to me that most considered discourse is not about reality but about representations of selected aspects of a particular domain of knowledge. Hence, models were a natural focus for my investigations, because they present facts belonging to one category in idioms of another—for example, representing music *as if* it were matter in motion. But when the two categories are identified as one, models function not as theories but as symbols of belief. There is, however, little agreement as to the main motives underlying the use of models,[4] so that, in addition to theory formation and symbolisation, models may serve a number of other functions, including 'simplification, reduction, extension, adequation, explanation, concretisation, globalisation, action or experimentation'.[5] In the case of non-formal sciences, for example, models have been introduced

> in function of relations between theories and theories, between experiments and theories, between experiments and experiments, between intellectual structures and the subjects using these structures, and in all these cases this has occurred in order dynamically to produce new results, or in order to tie up new ones with old ones as guarantees, or simply to establish relation.[6]

Model theory, however, is not the main theme of this book. That theme is the interrelatedness of knowledge, of which models are

---

[3] J. C. Kassler, *Inner Music: Hobbes, Hooke and North on Internal Character* (London, 1995).

[4] See, e.g., P. Ricoeur, *The Rule of Metaphor: Multi-disciplinary Studies of the Creation of Meaning in Language* tr. by R. Czerny et al. (London, 1978), pp. 239–245; J. Forge, 'Towards a theory of models in physical science', *Philosophy Research Archives* (1982), 8: 321–338; J. L. Aronson, R. Harré, and E. C. Way, *Realism Rescued: How Scientific Progress is Possible* (London, 1994), especially pp. 72–87.

[5] L. Apostel, 'Towards the formal study of models in the non-formal sciences', *The Concept and the Role of the Model in Mathematics and Natural and Social Sciences: Proceedings of the Colloquium sponsored by the Division of Philosophy of Sciences...*(Dordrecht, 1961), pp. 1–37, p. 3.

[6] Ibid.

prime instances, since in most, if not all, of their functions, establishing relations means cutting across lines that traditionally demarcate such separate domains as science and music. For example, science can find its models from the technology and semantic field of music, music can find its models from the technology and semantic field of science, and each domain may be guided by a philosophical or metaphysical principle. Thus, the title of this book is *Music, Science, Philosophy*. But the book itself is structured as a mirror image of its title. Part I, Philosophy (Chapters 1–3, and Part II, Science (Chapters 4–6), provide instances of music's contribution to epistemology, logic and the early modern sciences of developmental biology ('generation'), continuum mechanics, anatomy and physiological psychology, whereas Part III, Music (Chapters 7–10), provides instances of what some other domains have given back to the philosophy and theory of music. By way of preface, therefore, I shall summarise briefly the contents of the book, which proceed as a series of problems.

Part I begins with an epistemological problem as an instance of a continuing polemic in the West between proponents of two different kinds of knowledge. On the one hand, there are those who argue that knowledge derives from outside the human perceiver and, hence, is objective; on the other hand, there are those who insist that knowledge originates from within the human perceiver and is subjective. In the twentieth century the side of objective knowledge was taken by Sir Karl Popper, whose studies in music led him to formulate the three main principles of his philosophy of science. But Popper also applied these principles to a critique of the 'ideology' of Arnold Schoenberg and other modernist composers, an ideology that Popper supposed was subjectivist. Chapter 1 provides background for understanding Popper's polemic, which is critically examined.

In Chapter 2 the problem becomes: why did the development of combinatorial techniques, along with concepts of randomness and probability, emerge during the second half of the seventeenth century? The traditional history of science interpretation had failed to answer this question, so the philosopher of science, Ian Hacking, sought to offer a new historical interpretation, one that could explain the conjunction of mathematics and probability. Like the historians of science, Hacking associated the statistical or combinatorial aspect of modern probability with games of chance, thereby ignoring evidence from the philosophy, theory and practice of music. A consideration of this evidence suggests that the emergence of mathematical probability was facilitated by the impact of the Cartesian mechanical philosophy.

According to historians, that philosophy marks the beginning of 'the mechanisation of the world picture', which is best symbolised by clockwork.[7] But a more appropriate symbol might be the barrel of a music box or organ, since this device—a kind of proto-computer—mechanised the operations of encoding, storing and retrieving pieces of music. During the early modern period, logic, the branch of philosophy that treats forms of thinking in general, did not partake of the same kind of mechanisation. Yet, as Chapter 3 shows, the nineteenth-century polymath, William Stanley Jevons, established some general principles upon which computers are based, when he solved the problem of mechanising the operations of logical inference by recourse to the technology and semantic field of music.

Part II, on science, extricates the theoretical models of three early-modern natural philosophers. The first philosopher, William Harvey, is best known as the discoverer of the pathway of the blood—its circulating motion; but he also investigated 'the pathway to attaining form', namely, those generating motions or beginner events involved in making a new life. In Chapter 4, therefore, the problem is: how did Harvey find adequate concepts for describing techniques involved in making? Older concepts, inherited from the science of classical antiquity, were indebted to technical models from the visual arts. By their very nature such models were static and, hence, inadequate for describing beginner events. Adequate descriptions of such events require a dynamic model, which Harvey sought in the measures of dance.

Harvey's friend, Thomas Hobbes, also investigated motion. But this study became more urgent after 1647, the year he fell ill and, subsequently, developed a disorder of movement (tremor). To understand this disorder, Hobbes experimented with strings stretched over the soundboard of a musical instrument or the lath of a bow.

---

[7] See E. J. Dijksterhuis, *The Mechanization of the World Picture* tr. by C. Dikshoorn (Oxford, 1969), p. 415, a seminal study published first in Amsterdam in 1950.

From his observations and preliminary experiments on these analogues, Hobbes constructed a system of concepts that represent deformable media—the continuum mechanics that he presented in his treatise, *De corpore* (1655, tr. 1656). After a brief historical review of biomechanics, with special attention to forces that control body movement, Chapter 5 outlines the main features of Hobbes's theoretical model and treats the reception of his mechanics by two 'virtuosi'.

With Thomas Willis we are confronted with a different problem, one that derives from his belief that a person is both affected and not affected by a passion. Willis solved this paradox by distinguishing between intellect as the immortal musician, 'phantasie' as the musician's skill and body as the musician's hydraulic organ—a compound machine that operates on an hydro-pneumatic principle. Chapter 6 extricates the theoretical model behind Willis's anatomy and which allowed him to describe the person in meteorological terms. The normative person (intellect) is inspired by gentle, even winds that blow the organ regularly in order to ends; but if those winds are perturbed, the unregenerate human equivalent (phantasie) is 'troubled' by blasts that blow the organ irregularly.

In Part III the domain is music theory, which, in Chapter 7, is introduced as the science of music. By means of a thematic interpretation the introduction outlines the ancient and medieval conception of that science and charts its transition from ancient to modern musical science. The ancient and medieval doctrine of music is found chiefly in the old compendia of musical lore, which by the seventeenth century were being replaced by books of an increasingly specialised nature. The new books explicate and reduce to rules more and more divisions of musical knowledge, so that by 1830 nearly all the modern conceptual categories in music were distinguished and lines of theory and pedagogy established that have persisted well into the twentieth century.

Although the next two chapters fall within the domain of the history of music theory, they take an approach that is different not only from Chapter 7 but also from most previous histories. Older approaches tended to treat the history of music theory as a record of results, whereas the approach adopted in Chapters 8 and 9 conceives the history of music theory as a history of constructions of music-theoretic reality—that is, as representations of music. Chapter 8 treats the representation of an eighteenth-century music theorist, William Jones, whose construction is based on a variety of mechanics that is anti-Newtonian, whereas Chapter 9 focuses on the representation of an early twentieth-century music theorist, Heinrich Schenker, whose construction is based on a variety of developmental biology that is romantic.

Finally, we come to the question: what constraints are brought to the task of the acquisition, comprehension and use of musical knowledge? This question raises the problem of 'musical ear', solutions to which have been given chiefly by philosophers and by a variety of natural and social scientists, rather than by musicologists. Chapter 10 seeks to expose the problem as requiring also the investigations of musicologists, by showing how, historically and contemporaneously, solutions have been guided by two traditional theories of knowledge and learnability.

Nine of the ten chapters were published elsewhere, as indicated in the Acknowledgments. To the original versions I have introduced two kinds of alterations: (1) in most cases, titles are revised, and there are occasional minor emendations to the text or notes; and (2) in a few cases, there are major additions to, or abbreviations of, the original version for the purpose of providing greater cohesion to the book as a whole. Although as a collection the book does not present a single sustained argument, it still can be read in the sequence outlined above. But there are other ways of reading it as well.

For example, Chapters 1, 4, 6, 9 and 10 share a common problem of starting points, where something comes from something. Here, Popper (Chapter 1) and Schenker (Chapter 9) agree that the originative something is a preformed structure—for Popper, a myth, for Schenker, a primitive composition—from which grows a scientific theory and a musical composition, respectively. For Harvey (Chapter 4), however, the originative something is a liquid contained within a primordium. That liquid is a completely unstructured preconception, out of which a structured or formed conception emerges epigenetically. The problem of starting points reappears, though in a different guise, when considering the origin of 'musical ear' (Chapter 10) and when considering the issue of personal identity (Chapter 6).

In Chapters 2, 3, 5, 7 and 8 the shared problem relates to matter in motion. Stretched strings that vibrate are an instance of matter in motion, and experiments on these analogues provided Hobbes with a theoretical model for developing continuum mechanics (Chapter 5), one of the sciences of motion. About the same time, probability (Chapter 2) was conjoined with the mathematical art of combinations and permutations, which may be regarded as an idealisation of matter in motion. But mathematics is only one symbol system, and the idealisation of matter in motion may occur in such symbol systems as logic (Chapters 3) or music (Chapters 7, 8), where a sequence of letters or a sequence of notes may change their positions or arrangements in time (in music, for example, by augmentation, diminution, transposition, inversion or retrograde inversion).[8]

Readers, of course, will approach the collection with their own particular interests. But I hope the result of their reading will be a firm conviction that music's technology—its instruments and techniques—and music's semantic field—its music-theoretical constructs—have contributed in important ways to the growth of knowledge in science and in philosophy. This has been my own conviction and the impetus behind the kind of investigations included here.

[8] See also J. Wilkins, *Mercury: Or the secret and swift Messenger. Shewing, how a Man may with Privacy and Speed communicate his Thoughts to a Friend at any Distance* (London, 1641), where the author states that the 'secrecy of Speech...consists in a changing of the known Language, whether by Inversion, Transmuation, Diminution, Augmentation'.

ACKNOWLEDGMENTS

With the exception of Chapter 4, which began as a conference paper, the remaining nine chapters originated either as an article in a journal or as an essay in, or introduction to, a book. Some of the investigations leading to publication were supported by the Australian Research Council, and all previously published material is included here with the requisite permissions as indicated in the chronological list below.

'The systematic writings on music of William Jones (1726–1800), *Journal of the American Musicological Society* (1973), 26: 92–106. © 1973 by the American Musicological Society. All rights reserved.

'The "science" of music to 1830', *Archives Internationales d'Histoire des Sciences* (1980), 30: 111–136, an abbreviated version of the introduction to J. C. Kassler, *The Science of Music in Britain, 1714–1830: A Catalogue of Writings, Lectures and Inventions* (2 vols., New York: Garland Publishing, Inc., 1979).

'Heinrich Schenker's epistemology and philosophy of music: An essay in the relations between evolutionary theory and music theory', *The Wider Domain of Evolutionary Thought* ed. by D. Oldroyd and I. Langham [Australasian Studies in History and Philosophy of Science Vol. 2]. Copyright © 1983 by D. Reidel Publishing Co., Dordrecht, Holland, pp. 221–260. Used by kind permission of Kluwer Academic Publishers.

'Apollo and Dionysos: Music theory and the western tradition of epistemology', *Music and Civilization: Essays in Honor of Paul Henry Lang*, ed. by Edmund Strainchamps et al. Copyright © 1984 by Edmond Strainchamps and Maria Rika Maniates, pp. 457–471. Used by permission of W. W. Norton & Company, Inc., New York.

'The emergence of probability reconsidered', *Archives Internationales d'Histoire des Sciences* (1986), 36: 17–44.

'William Stanley Jevons: Music and the mechanisation of logic', *Australasian Music Research* (1996), 1: 327–335.

'Restraining the passions: Hydropneumatics and hierarchy in the philosophy of Thomas Willis', *The Soft Underbelly of Reason: The Passions in the Seventeenth and Eighteenth Centuries* ed. by S. Gaukroger [Routledge Studies in Seventeenth-Century Philosophy, Vol. 1], London: Routledge, 1998, pp. 147–164.

'Musicology and the problem of musical ear', *Musicology Australia* (1999), 22: 18–30.

'The ovum as initial datum: William Harvey's theory of creativity', *Conference on Early Modern Thought, 1543–1794*, The University of Sydney, 1999.

'On the stretch: Hobbes, mechanics and the shaking palsy', *1543 and All That: Word and Image in the Proto-scientific Revolution* ed. by G. Freeland and A. Corones [Australasian Studies in History and Philosophy of Science Vol. 13], pp. 151–187. Copyright © 2000 by Kluwer Academic Publishers, Dordrecht, The Netherlands. Used by kind permission of Kluwer Academic Publishers.

The life of a scholar can be lonely, especially if one has no permanent 'home' within a particular discipline or is isolated in other ways. Kindred spirits, even those whom one never meets except through correspondence, thus form an important component of the scholarly life. For their guidance, encouragement or shared enjoyment of the subject matter, a few kindred spirits deserve special mention: Robert Crane as teacher of musical composition, Peter E. de Janosi as general mentor, and Paolo Gozza as ideal reader. Of course, numerous other persons have been helpful over the course of nearly thirty years, most of whom were acknowledged in previous publications, including those listed above. A debt remains to Janet D. Hine, for guiding my preparation of the index; to Penelope Gouk, for suggesting a collection of some of my writings, and to Roy Porter, for publishing my first articles in the history of science, even after a referee of one of those articles asked: 'what does music have to do with science?'.

1. APOLLO OR DIONYSUS?
MUSIC AND THE GROWTH OF KNOWLEDGE

[Divine madness is] the source of man's appreciation of rhythm and harmony, and Apollo and the Muses and Dionysus were the gods who cooperated to implant it in us.
Plato, *Laws* (627c8–d3)

The beginnings of music theory are to be found deep in western mythology, where an antithesis was established between order and disorder. Order was the domain of Apollo, who radiated multiple functions from his godhead. As preserver of the harmony of life, Apollo purged the soul of man from guilt and cleansed the body of its ills, thus becoming the god of purification (*katharsis*). His regular movement through the heavens produced order in the universe by imparting harmony and rhythm of form; wherefore, he was not only leader of the Muses but also god of poetry, dancing and music. His widespread legal activity embraced everything in religious and secular life that could be brought under rules and regulations, and so, as god of law and order, his maxims enjoined limit, moderation and obedience to authority. But most important, among those Greek divinities that were personifications of cosmic forces, Apollo symbolised rational powers and represented pure intellect.[1]

Apollo was a Greek god; but his antithesis, Dionysus, was a foreign god.[2] The many different names for Dionysus outside of

---

[1] For the various attributes of Apollo, see W. K. C. Guthrie, *The Greeks and their Gods* (London, 1950), pp. 183–204; see also B. Meinecke, 'Music and medicine in classical antiquity', *Music and Medicine* ed. D. M. Schullian and M. Schoen (Freeport, 1971), pp. 47–52.

[2] Guthrie, ibid.

Greece suggest that he stood for the 'wet element' in nature—wine, the juicy sap of plants, the life blood of animals, semen and, by extension, creative powers.[3] By the time his cult was introduced into Greece, however, disorder was the domain of Dionysus, who symbolised the cosmic forces of irrationalism and represented sensuous appetites. His rites were intended to lead to a final state of ecstasy (*ekstasis*, standing outside oneself) and divine possession (*enthousiasmos*).[4] Indeed, the most characteristic parts of the Dionysian *orgia* were music and dancing as aids to ecstasy, to the sense of union with the godhead and to the power of seeing visions.

The intelligible, determinate and mensurable domain of Apollo's order, then, was opposed to the fantastic, vague and shapeless domain of Dionysus' disorder. These two domains were further distinguished by the music and musical instruments associated with each.[5] On the one hand, Apollo's concordant music quelled the passions; and his instrument, the kithara or Greek lyre, was the model of proportional tuning systems based on rational numbers. On the other hand, Dionysus' dissonant and 'barbarous' music raised the passions; and his instrument, the aulos or Greek oboe, produced its sounds by air pressure and, hence, required tempered systems based on irrational numbers.[6]

The antithesis of Apollo and his kithara and Dionysus and his aulos formed the basis of the Greek theory of the soul, for the Greeks held that, to achieve its purpose, the soul had been endowed with a dual nature: a rational nature, concerning itself with reason conceived as a divine, incorporeal element; and an irrational nature, concerning

---

[3] Ibid., pp. 174 and 181, who states that the most common name for Dionysus—Bacchus—denoted both the deity and the devotee.

[4] Ibid., pp. 147ff.; see also Euripides, *Bacchae* ed. and tr. by E. R. Dodds (Oxford, 1960).

[5] W. D. Anderson, *Ethos and Education in Greek Music: The Evidence of Poetry and Philosophy* (Cambridge, Massachusetts, 1966), pp. 7–10 *et passim*.

[6] For the tuning problem, see J. W. McKinnon, 'Aulos', *The New Grove Dictionary of Music and Musicians* ed. by S. Sadie (20 vols., London, 1980), vol. 1, pp. 699–702; see also J. C. Kassler, 'Music as a model in early science', *History of Science* (1982), 20: 103–139.

itself with the corporeal body through sensuous desire.[7] This twofold division—with its hierarchy of higher (rational) and lower (irrational) parts—was retained not only in most western theories of knowledge but also in most western theories of music after myth began to give way to science, for two different philosophies were established, the one, rationalism, based on the intellect, the other, empiricism, based on the senses. The two types of knowledge resulting were then distinguished as objective and subjective knowledge.[8]

Pythagoras and his followers stressed the objective nature of musical knowledge, since, for them, the principles of music are founded in certain general and universal laws, into which all that we discover in the material world of harmony, symmetry, proportion and order are resolvable.[9] That is, it was supposed that music's principles, like the principles of all things, were founded in mathematical truth and resulted from some general and universal laws of nature—laws, of course, that were conceived as divinely ordained. Therefore, the excellence of music was thought to be intrinsic, absolute and inherent and, in short, resolvable into number, weight and measure.[10] This

---

[7] See, e.g., K. R. Popper and J. C. Eccles, 'The mind-body problem in Greek philosophy', *The Self and its Brain* (Berlin, 1977), pp. 159–71; see also G. S. Kirk, J. E. Raven and M. Schofield, *The Presocratic Philosophers* (2d edn., Cambridge, 1983), p. 21.

[8] Insofar as rationalism dominated western thought until about the end of the sixteenth century, empirical evidence was mustered only occasionally and only to verify or demonstrate *a priori* propositions. Moreover, the sway of rationalism led to the study and development of the logic associated with it—deductive logic. Only after 1600, when empiricism began to assert itself, were investigations undertaken into the nature of its particular logic—induction. This change coincided with the rise of psychology and aesthetics as specialised disciplines.

[9] The main Pythagorean texts on music are the Euclidean *Sectio canonis* and Ptolemy's *Harmonika*, for which see T. J. Mathieson, 'An annotated translation of Euclid's Division of the Monochord', *Journal of Music Theory* (1975), 19: 236–258, and I. Düring, *Die Harmonielehre des Klaudios Ptolemaios* (Göteborg, 1930); for other Pythagorean texts, see A. Barker, *Greek Musical Writings*...(2 vols., Cambridge, 1984-89); see also E. A. Lippman, *Musical Thought in Ancient Greece* (New York, 1964).

[10] For number, weight and measure, see Kassler, 'Music as a model' (note 6), pp. 126–127. As arithmetic, geometric and harmonic proportion, these were the principal formal categories of early science. Subsequently, these categories

hypothesis, which is the basis of most formalist theories of music, was adopted by the seventeenth-century physicist, Robert Hooke, and by the eighteenth-century mathematician, Leonhard Euler, among others.

In contrast to the objective approach of the Pythagoreans, there was the subjective theory of Theophrastus, which was developed into a systematic theory of music by Aristoxenus.[11] Theophrastus argued that hearing is the sense that most deeply stirs our emotions and that music is a movement of the soul whereby we are freed of the evils ensuing from passion. But the soul is moved by divinity, not by music, for Theophrastus believed that what a musician composes is not his own work but rather a message or dispensation from the gods, especially the Muses. The musician is thus the instrument through which the Muses speak—a mouthpiece of the gods: possessed by a divine spirit, he becomes frenzied and communicates this state to the audience.[12] For Theophrastus, then, the first cause of music is divinity, and the second cause is emotion, namely, inspiration, pleasure or pain. Any of these emotions may so change the character of the voice that it loses its customary form and becomes musical. This hypothesis, which is the basis of most expressionist theories of music, was adopted by the eighteenth-century philosopher, Jean-Jacques Rousseau, and by the nineteenth-century psychologist, Herbert Spencer, among others.

If we take the theory of inspiration and frenzy and discard its divine source, we arrive at a theory of art as self-expression or, more precisely, as self-inspiration and the expression and communication of emotion. This modification replaces the gods or God with the

---

underwent a change in meaning to signify experimentally numbered, weighed and measured, so that new formal categories were constructed, as in the Cartesian philosophy, with its categories of number, motion and extension, or in the Newtonian, with its categories of mass, length and time.

[11] F. Wimmer, *Theophrasti Eresii Opera...omnia* (Paris, 1866); H.S. Macran, *The Harmonics of Aristoxenus* (Oxford, 1902); L. Rowell, 'Aristoxenus on rhythm', *Journal of Music Theory* (1979), 23: 63–79; Aristoxenus, *Elementa rhythmica: The Fragment of Book II and the Additional Evidence*...ed. and tr. with a commentary by L. Pearson (Oxford, 1990).

[12] This process of communication was compared either to sympathetic resonance or to magnetism, occult phenomena for which manifest explanations were sought from about 1600 onwards.

hidden nature or the essence of the artist. Hence, since the artist is thought to inspire himself and since the source of inspiration is unconscious and therefore irrational, the theory is open to the charge of subjectivism, the very charge made by Sir Karl Popper, an eminent philosopher of science.[13] Born in Vienna in 1902, Popper left Austria in 1937 at the time of the Nazi ascendancy to take up a senior lectureship at the then Canterbury University College, Christchurch, New Zealand. Subsequently, he made his home in England, where in 1945 he became reader and in 1949 professor of logic and scientific method at the London School of Economics.

According to Popper, questions like 'What is art?' are never genuine questions, since they lead to the theory of art as self-inspiration. During his student days in Vienna, this subjectivist theory re-emerged as the so-called 'expressionist' movement; and one of its most prominent contributors, the composer, Arnold Schoenberg, provided a succinct statement of the movement's theory, when he wrote:

> a work of art can produce no greater effect than when it transmits the emotions which raged in the creator to the listener, in such a way that they also rage and storm in him. ... [Indeed,] there is only one greatest goal towards which the artist strives: *to express himself.* If that succeeds, then the artist has achieved the greatest possible success; next to that, everything else is unimportant... .[14]

For Popper, such a theory is an empty theory, because everything a man or animal can do is, among other things, an expression of an internal state, of emotions or of personality. Since this holds trivially true for all kinds of human and animal 'languages', the important question to ask about art is 'what makes a work interesting or significant?' That is, the focus should be on the work of art itself, since only in this way can an objectivist theory of art be achieved.

---

[13] Popper died on 17 September 1994.

[14] A. Schoenberg, *Style and Idea: Selected Writings* ed. by L. Stein and tr. by L. Black (London, 1975), pp. 450 and 454. Popper ignores the religious motivation behind modern expressionism, but Schoenberg's writings indicate that he never propounded a 'theology without God'.

Popper does not deny self-expression in art, but he stresses its utter triviality. For him, the really interesting function of an artist's emotions is not that they are to be expressed but that they may be used to test the success or fittingness of the impact made by the objective work. The artist may use himself as a kind of test body, and he may modify his artwork or even discard it whenever he is dissatisfied by his own reaction to it. Popper, therefore, adumbrates a theory of art not as the product of artistic inspiration but as the result of an artist's self-criticism. According to him, such a theory would be objective, for it is the artwork itself that is responsible for the emotions of the artist rather than the other way round.

Now, Popper came to these conclusions through his interest in, and knowledge of, music. Indeed, scholars have neglected the importance of music in Popper's thought; and it is not widely known that for a time he was a student of music. First, he studied musical composition with Erwin Stein, a pupil of Schoenberg; then, he became a student of church music in the Vienna Conservatory; and finally, he chose the study of the history of music as a second subject for his doctoral examination at the University of Vienna. According to Popper,

> out of my interest in music there came at least three ideas which influenced me for life. One was closely connected with my ideas on dogmatic and critical thinking, and with the significance of dogmas and traditions. The second was a distinction between two kinds of musical composition, which I then felt to be immensely important, and for which I appropriated for my own use the terms 'objective' and 'subjective'. The third was a realization of the intellectual poverty and destructive power of historicist ideas in music and in the arts in general.[15]

Regarding the first idea, Popper, during his study of music, arrived at what he recognised as 'perhaps untenable historical conjecture', namely, that musical and scientific creation seem to have this much in common:

---

[15] K. R. Popper, *Unended Quest: An Intellectual Autobiography* (Glasgow, 1976), p. 55.

> the use of dogma, or myth, as a man-made path along which we move into the unknown, exploring the world, both creating regularities or rules and probing for existing regularities. And once we have found or erected, some landmarks, we proceed by trying new ways of ordering the world, new coordinates, new modes of exploration and creation, new ways of building a new world, undreamt of in antiquity unless in the myth of the music of the spheres.
>
> Indeed, a great work of music (like a great scientific theory) is a cosmos imposed upon chaos—in its tension and harmonies, inexhaustible even for its creator. This was described with marvelous insight by Kepler in a passage devoted to the music of the heavens... .[16]

This historical conjecture was reinforced by Popper's reading of Immanual Kant's *Critique of Pure Reason*, in which the central idea is that '*scientific theories are man-made, and that we try to impose them upon the world*'.[17] Combining Kant's insight with some ideas of his own, Popper concluded:

> Our theories, beginning with primitive myths and evolving into the theories of science, are indeed man-made, as Kant said. We do try to impose them on the world, and we can always stick to them dogmatically if we so wish, even if they are false (as are not only most religious myths, it seems, but also Newton's theory, which is the one Kant had in mind). But although at first we have to stick to our theories—*without theories we cannot even begin*, for we have nothing else to go by—we can, in the course of time, adopt a more critical attitude towards them. We can try to replace them by something better if we have learned, with their help, where they let us down. Thus there may arise a scientific or critical phase of thinking, *which is necessarily preceded by an uncritical phase*.[18]

Dogmatic thinking, which Popper regarded as pre-scientific, is a necessary stage if critical thinking is to be possible, because critical thought must have before it something to criticise, and this something, according to Popper, must result from dogmatic thinking.

---

[16] Ibid., pp. 58–59.

[17] Ibid., p. 59.

[18] Ibid., pp. 59–60.

When applied to the growth of knowledge, therefore, the problem becomes: how to demarcate science from pseudo-science or myth; and Popper's attempt to solve this problem also derived from his interest in music. According to his own account, he was early struck by what he thought was a difference between the compositions of Bach and those of Beethoven. This difference, he supposed, was the product of divergent attitudes, Bach's being objective and Beethoven's subjective.[19] The accuracy of Popper's characterisations is not germane here. What interests us is his distinction between an objective and a subjective approach or attitude, since this antithesis became decisive in Popper's theory of knowledge.

In that theory Popper divides our knowledge of the universe into three 'worlds': world 1, the physical world, consists of material things; world 2, the psychological world, consists of subjective thoughts in the minds of living creatures; and world 3, the creative world, consists of objective structures that are products of minds. Thus, world 3 contains our cultural heritage, and its objects include such things as works of art and scientific theories. But world 3 comprises

> not only the products of our intellect, together with the unintended consequences which emerge from them, but also the products of our mind in a much wider sense; for example, the products of our imagination. Even theories, products of our intellect, result from the criticism of myths, which are products of our imagination: they would not be possible without myths; nor would criticism be possible without the discovery of the distinction between fact and fiction, or truth and falsity. This is why myths and fictions should not be excluded from world 3. So we are led to include art and, in fact, all human products into which we have injected some of our ideas, and which incorporate the result of *criticism* (in a sense wider than merely intellectual criticism). We ourselves may be included, since we absorb and criticise the ideas of our predecessors, and try to form ourselves; and so may our children and pupils, our traditions and institutions, our ways of life, our purposes, and our aims.[20]

---

[19] Ibid., pp. 60–68.

[20] Ibid., p. 195.

In addition to 'embodied' objects such as books and paintings, world 3 includes much that is 'unembodied', as is clear from Popper's treatment of the status of a musical composition. This object, he asserts,

> has a very strange sort of existence. Certainly it at first exists encoded in the musician's head, but it will probably not even exist there as a totality, but, rather, as a sequence of efforts or attempts; and whether the composer does or does not retain a total score of the composition in his memory is in a sense not really essential to the question of the existence of the composition once it has been written down. But the written-down encoding is not identical with the composition—say, a symphony. For the symphony is something acoustic and the written-down encoding is obviously merely conventionally and arbitrarily related to the acoustic ideas which this written-down encoding tries to incorporate and to bring into a more stable and lasting form. So here there already arises a problem. Let us pose the problem in the following way. Clearly, Mozart's Jupiter Symphony is neither the score he wrote, which is only a kind of conventional and arbitrarily coded statement of the symphony; nor is it the sum total of the imagined acoustic experiences Mozart had while writing the symphony. Nor is it any of the performances. Nor is it all performances together, nor the class of all possible performances. This is seen from the fact that performances may be good or less good, but that no performance can really be described as ideal. In a way, the symphony is the thing which can be interpreted in performances—it is something which has the possibility of being interpreted in performance. One may even say that the whole depth of this World 3 object cannot be captured by any single performance, but only by hearing it again and again, in different interpretations. In that sense the World 3 object is a real ideal object which exists, but exists nowhere, and whose existence is somehow the potentiality of its being reinterpreted by human minds. So it is first the work of a human mind or of human minds, the product of human minds; and secondly it is endowed with the potentiality of being recaptured, perhaps only partly, by human minds again.[21]

From this conception of our created world—of the objective knowledge contained in world 3, Popper mounts his attack on

---

[21] Popper and Eccles, *The Self and its Brain* (note 7), pp. 449–450.

subjectivist theories and, in particular, on psychological theories of knowledge and learning by induction. Popper claims there is no such thing as unprejudiced observation, since all observation is an activity with an aim. He also claims that there is no such thing as passive experience, since experience is the result of active exploration by the organism in the course of its search for regularities or invariants. And he further claims that there is no such thing as perception without the context of interests and expectations and, hence, of regularities or 'laws'. As a consequence of these claims, Popper argues that knowledge and learning take place in a hypothetico-deductive manner and that

> conjecture or hypotheses must come before observation or perception: we have inborn expectations; we have latent inborn knowledge, in the form of latent expectations, to be activated by stimuli to which we act as a rule while engaged in active exploration. All learning is modification (it may be refutation) of some prior knowledge and thus in the last analysis of inborn knowledge.[22]

Since Popper maintains there can be 'knowledge without a knowing subject', he considers subjectivist approaches and theories of knowledge to be utterly misconceived, because they overlook a vitally important aspect of knowledge in limiting knowledge to the knowing subject. Such theories, according to Popper, treat world-3 objects as if they were 'mere utterances or expressions of the knowing subject'. Popper attempts to replace this approach by a very different one:

> Admitting that world 3 originates with us, I stress its considerable autonomy, and its immeasurable repercussions on us. Our minds, our selves, cannot exist without it; they are anchored in world 3. We owe to the interaction with world 3 our rationality, the practice of critical and self-critical thinking and acting. We owe to it our mental growth. And we owe to it our relation to our task, to our work, and its repercussions upon ourselves ... It is through the attempt to see objectively the work we have done—that is to see it critically—and to

[22] Popper, *Unended Quest* (note 15), p. 52.

> do it better, through the interaction between our actions and their objective results, that we can transcend our talents, and ourselves.[23]

Dialectic, then, is fundamental to Popper's thinking, for he argues that the critical process alone contributes to the growth of knowledge. And he further asserts that only through the objective knowledge contained in world 3 is it possible for ideas to have histories.

This assertion brings us to the third and last position adopted by Popper through his studies in music, namely, the poverty of historicism.[24] By the term 'historicism', Popper denotes any argument from design—that is, any doctrine that there are general laws of development rendering the course of history inevitable and predictable.[25] As opposed to arguments from design, Popper propounds an open-ended evolutionism in which the end or goal of history is not predictable and in which the course of events may be freely influenced by human creativity. He argues, therefore, that

> the universe, or its evolution, is creative, and ... the evolution of sentient animals with conscious experiences has brought about something new. These experiences were first of a more rudimentary and later of a higher kind; and in the end that kind of consciousness of self and that kind of creativity emerged which, I suggest, we find in man.
>
> What I describe by the word 'creative' is described by Jacques Monod ... when he speaks of the unpredictability of the various species, and especially our own human species... .[26]

---

[23] Ibid., p. 196.

[24] Ibid..

[25] K. R. Popper, *The Poverty of Historicism* (London, 1979). The dedication—'In memory of the countless men and women of all creeds or nations or races who fell victims to the fascist and communist belief in Inexorable Laws of Historical Destiny'—reveals the fundamental thesis of his book, p. iv, 'that the belief in historical destiny is sheer superstition, ... that there can be no prediction of the course of human history by scientific or any other rational methods'.

[26] Ibid., where Popper singles out the panpsychism of Plato and Hegel, the materialism of Marx and other historical methods modelled on classical (deterministic) physics. See also K. R. Popper, *The Open Society and its Enemies* (London, 1945).

For Popper, the critical process alone provides the impetus for the evolution of ideas, and the kind of critical process he advocates—'falsification'—constitutes the essence of his methodology of science, namely, a process of trial-and-error elimination or 'conjecture and refutation'.[27] Consisting generally 'in the search for difficulties or contradictions and their tentative solutions',[28] Popper's critical process commences with an act of the imagination—the proposal of bold hypotheses. We then expose these hypotheses to the severest criticism in order to detect where we have erred. In other words, we start our investigation with a problem, the solution to which, always tentative, entails an imaginative conjecture, a hypothesis or theory. Various competing theories are then compared and discussed critically, so that we may detect their shortcomings. And the ever-changing, ever-inconclusive results of such critical debate constitute what Popper calls 'the science of the day'. For Popper, theories are our own inventions, but these theories 'may be merely ill-reasoned guesses, bold conjectures, hypotheses'; out of these 'we create a world, not the real world, but our own nets in which we try to catch the real world'.[29]

Popper's methodology of science has attracted wide attention among philosophers of science as well as among some scientists. Little notice, however, has been taken of its relation to art. Yet, Popper not only attacks the long-lived theory of artistic inspiration

---

[27] Popper and Eccles, *The Self and its Brain* (note 6), pp. 15–16. Popper's reference is to J. Monod, *Chance and Necessity* (Glasgow, 1977). On the subject of creative, i.e., 'emergent', evolution, Popper has this to say, p. 17:

> Against the acceptance of the view of emergent evolution there is a strong intuitive prejudice. It is the intuition that, if the universe consists of atoms or elementary particles, so that all things are structures of such particles, then every event in the universe ought to be explicable, and in principle predictable, in terms of *particle structure* and of *particle interaction.*

Popper's creative evolutionism is based on the Darwinian principle of natural selection, but it is worth noting that other versions of creative evolutionism draw on arguments from design. For one example, see Chapter 9 *infra*.

[28] K. R. Popper, *Conjectures and Refutations: The Growth of Scientific Knowledge* (London, 1978); see also Popper and Eccles, *The Self and its Brain* (note 6), pp. 171–176.

[29] Popper , *Unended Quest* (note 15), p. 60.

but also consistently decries what he believes to be its concomitant historicist element. The latter, in Popper's opinion, is a consciousness of novelty, of having to produce something new.[30] Such a stance, according to Popper, has nothing to do with the fundamental aim of producing artworks that are the best of their kind. Moreover, he wonders why we ought not to produce old-fashioned music in the style, say, of Bach, Haydn or Mozart. The rage for novelty that Popper detects in twentieth-century music is, in his words, 'inspired by a mad religion, namely the religion of progress, the religion of hastening along as fast as possible and constantly looking back to see how far or how fast one goes'.[31]

Popper admits he is conservative in his musical tastes. Even though he was involved for a short time in the Society for the Private Performance of Music, which was founded chiefly to present compositions by Schoenberg and his pupils, Popper tells us that he 'turned in disgust from [the] modernism' of Schoenberg, 'which involved constantly observing oneself, to see whether one was going sufficiently fast beyond what had just been reached'[32] And about Schoenberg's pupil, Anton Webern, Popper had this to say: 'Nobody could doubt the purity of his heart', but 'there was not much music to

---

[30] Nonetheless, Popper allows for novelty in his evolutionary scheme, as is clear from his remarks in Popper and Eccles, *The Self and its Brain* (note 7), p. 14:

> The view that there is no new thing under the sun is, in a way, involved in the original meaning of the word 'evolution': to evolve means to unroll; and evolution meant originally the unrolling of what is there already: what is there, *preformed*, is to be made manifest. (To develop, similarly, means to unfold what is there.) This original meaning … still seems to play its role in the world-view of some materialists or physicalists.
>
> Today some of us have learnt to use the term 'evolution' differently. For we think that evolution—the evolution of the universe, and especially the evolution of life on earth—has produced new things: *real novelty* ... We should be more clearly aware of this real novelty.

[31] See the discussions at the end of each session of the Third C. H. Boehringer Sohn Symposium; see *Structure in Science and Art*...ed. by P. B. Medawar and J. H. Shelley (Amsterdam, 1980), p. 121. The music participants included Alexander Goehr, György Ligeti and Charles Rosen.

[32] Ibid., p. 122.

be found in his modest compositions'.[33] These comments appear to be the result of what Popper calls 'dogmatic thinking'; that is, they are pre-scientific statements awaiting criticism.[34]

Popper's published writings, at least, support this conclusion, for there is no evidence in them that he has examined, for example, either the compositions or the writings of Schoenberg, who, for all his emphasis on music as the expression of the inner life of the composer, shares with Popper a number of positions, including a commitment to the quest for knowledge,[35] an emphasis on human creativity,[36] an affirmation of a realist philosophy of interaction,[37] and a belief in an

---

[33] Popper, *Unended Quest* (note 15), p. 74.

[34] The critical retorts of Ligeti and Rosen (see note 31) do not constitute a systematic refutation of Popper's theory of art.

[35] See, e.g., A. Schoenberg, *Theory of Harmony* [1911] tr. by R. E. Carter (London, 1978), p. 1: what matters is 'the search itself'.

[36] Ibid., p. 319:

> The creative spirit strives for more, more and more; those who merely seek enjoyment are satisfied with fewer. Between this More and this Fewer the battles of art are fought. Here, truth, the search—there, aesthetics, that which has presumably been found, the reduction of what is worth striving for to what is within reach. What is worth striving for is to discover everything that lies within the natural tone, to attain thereby everything of which the human brain with its powers of association and its ability to systematize is capable. What is within reach has its temporary boundaries wherever our nature and the instruments we have invented have their temporary boundaries. What is attainable with the phenomenon outside ourselves, as far as the tone itself is concerned, theoretically speaking, has no boundaries. What has not yet been attained is what is worth striving for.

It should be noted that Popper and Schoenberg differ on the source of creativity: for the former, it is the province of scientific genius and for the latter, of artistic genius.

[37] Ibid., p. 30, who argued that

> order, ... clarity ... are there by chance, not by law, not by necessity; and what we claim to perceive as laws [defining order and clarity] may perhaps only be laws governing our perception, without therefore being the laws a work of art must obey. And that we think we see [laws, order] in a work of art can be analogous to our thinking we see ourselves in the mirror, although we are of course not there. The work of art is capable of mirroring what we project into it. The conditions our conceptual power imposes, a mirror image of our own nature ... , may be observed in the

open-ended evolutionism.[38] Yet, Popper singles out Schoenberg's project of expressionism, a project he dislikes for two principal reasons. First, according to the theory of inspiration, an artist is not responsible for his actions, since his mysterious 'gifts', indeed his whole personality, determine what he does. The result may be good or bad, according to whether an artist is gifted or merely talented. Carried to its extreme, therefore, the theory of inspiration leads to voluntarism and ethical relativism. As Schoenberg himself wrote:

> The artist's creative activity is instinctive. Consciousness has little influence on it. He feels as if what he does were dictated to him. As if he did it only according to the will of some power or other within him, whose laws he does not know. He is merely the instrument of a will hidden from him, of instinct, of his unconscious. Whether it is new or old, good or bad, beautiful or ugly, he does not know. He feels only the instinctual compulsion, which he must obey. And in this instinct the old may find expression, and the new. Such as depends on the past, and such as points out paths to the future. Old truths or new errors. [It is] his musical nature, as he inherited it from a musical ancestor or acquired it through the literature, but [it is] perhaps also the outflow of an energy that is seeking new paths. Right or wrong,

work. This mirror image does not, however, reveal the plan upon which the work itself is oriented, but rather the way we orient ourselves to the work. Now if the work bears the same relation to its author, if it mirrors what he projects into it, then the laws *he* thinks he perceives may also be just such as were present in his imagination, but not such as are inherent in his work. And what he has to say about his formal purposes could be relatively inconsequential. It is perhaps subjectively, but not necessarily objectively, correct. One has only to look in the mirror from another point of view; then one can believe that mirror image, too, is actually projected by the observer, only this time the image is different. Now even if one can assume with certainty that the observer will not see in the work of art something entirely different from what is actually in it—since object and subject do indeed interact—even so the possibility of misapprehension is still too great to allow us to say with absolute confidence that the presumed order is not just that of the subject. All the same, the state of the observer can be ascertained from the order he sees.

[38] Ibid., p. 31, where Schoenberg assumed there is only one 'eternal law', namely, 'evolution and change—this way has to be more fruitful than the other, where one assumes an end of evolution because one can thus round off the system'.

> new or old, beautiful or ugly—how does one know who only senses the instinctual urge? Who would dare to differentiate right from wrong in the instinct, in the unconscious, to keep separate the knowledge inherited from predecessors and the intuitive power granted by the spirit?[39]

Second—Popper's other reason, the theory of inspiration is rooted in irrationalism. And here, perhaps, is the crux of the problem, for the concern of philosophers traditionally has been seen as a search for the intelligible and the universal—what Popper calls 'invariants', 'regularities', 'laws'. This search, which is Apollonian, relies on conscious, sober, rational activity and on discursive notions of truth. In contrast, the concern of artists traditionally has been conceived as a search for the sensuous and particular—the expression of individual and personal emotions, especially the 'feeling' for form. This search, which is Dionysian, relies on unconscious, irrational forces and on non-discursive notions of truth. For philosophers, therefore, the artist's vision of truth is suspect, since this vision is thought to be apprehended in a non-sober and, hence, 'mad' intuition.

The classic statement of this kind of truth is found in two dialogues by Plato, *Phaedrus* and *Ion,* in which truth denotes not factual correctness but rather divine truth.[40] As Socrates puts it in the first dialogue (244a–b and 245a): 'the greatest blessings come by way of madness, indeed of madness that is heaven-sent'. Thus, 'if any man comes to the gates of poetry without the madness of the Muses, persuaded that skill alone will make him a good poet, then shall he and his works of sanity with him be brought to nought by the poetry of madness, and behold, their place is nowhere to be found'.[41]

For Plato, divine truth was none other than the ordering principle of all reality. Among the gods representing this divine truth,

---

[39] Ibid., p. 416.

[40] Plato, *The Collected Dialogues* ed. by E. Hamilton and H. Cairns (Princeton, 1973), pp. 215–228 and 475–525. See also M. H. Partee, 'Inspiration in the aesthetics of Plato', *Journal of Aesthetics and Art Criticism* (1971), 30: 87–95; K. Dorter, 'The Ion: Plato's characterization of art', *Journal of Aesthetics and Art Criticism* (1973), 32: 65–78; and Popper, *Unended Quest* (note 15), pp. 65–67.

[41] Plato, ibid., pp. 491–492.

it is the function of the Muses to infuse the artist with the ordering principles and, through him, infuse all who participate in the work he has created. It therefore follows that the artist has little or no conscious control over his art, a point made in the second dialogue (533d–534a), where Socrates points out to the rhapsode, Ion, that

> this gift you have of speaking well on Homer is not an art [i.e., *technē*]; it is a power divine, impelling you like the power in the ... magnet ... This stone does not simply attract the iron rings, just by themselves; it also imparts to the rings a force enabling them to do the same thing as the stone itself, that is, to attract another ring, so that sometimes a chain is formed, quite a long one, of iron rings, suspended one from another. For all of them, however, their power depends upon the loadstone. Just so the Muse. She first makes men inspired and then through these inspired ones others share in the enthusiasm, and a chain is formed, for the epic poets, all the good ones, have their excellence, not from art, but are inspired, possessed, and thus they utter all these admirable poems. So it is also with the good lyric poets; as the worshipping Corybantes are not in their senses when they dance, so the lyric poets are not in their senses when they make these lovely lyric poems. No, when once they launch into harmony and rhythm, they are seized with the Bacchic transport, and are possessed—as the bacchants, when possessed, draw milk and honey from the rivers, but not when in their senses.[42]

Plato recognised that art is capable of revealing the intelligible and universal in the beauty of its structure, that is, through art's 'harmony' and 'rhythm'. These principles have cosmic significance for Plato, since he regarded them as reflections of divine truth. Imitation of divine truth furnishes art with its harmony and rhythm, and it is here that the distinctiveness of art lies and here also that we may seek its formal, Apollonian properties. But imitation is the result not of inspiration, the downward movement from the gods to the artist, but rather of aspiration, the upward movement from the artist to the gods. And Plato conceived this upward motion as a tendency of

[42] Ibid., pp. 219–220. According to Guthrie, *The Greeks* (note 1), p. 154, the Corybantes were servants of a cult of ecstatic or orgiastic ritual; and the Greek verb *korybantiaein* meant to be in a state of divine madness such that hallucinations occurred. Greek medical writers identified this state as a pathological condition.

the mind to ascend toward recollecting, however faint, divine truth. Thus, 'the function of the Muses, who provide man with the power of producing beauty, may be seen to be connected with philosophy'.[43]

Plato's two kinds of movement, inspiration and aspiration, are connected with two kinds of prophecy: the prophecy of the mystic, for whom intuition is decisive, and the prophecy of the augur, who has learned the art (*technē*, know-how, skill) of interpreting signs of the intelligible and universal. We may consider Schoenberg and Popper, respectively, as each kind of prophet insofar as both men believe in an objective and discernible truth. Yet, for Schoenberg, as for Popper, no man is ever in certain possession of that truth.[44] Despite these shared assumptions, Popper distrusts Schoenberg's

---

[43] Dorter, 'The Ion', (note 40), p. 76.

[44] In *Conjectures and Refutations* (note 28), p. 376, Popper states that one can appreciate 'every step, every approach, … as valuable, indeed as invaluable', that 'our traditions often help to encourage such steps' and that 'without an intellectual tradition the individual could hardly take a single step towards the truth'. In short, knowledge is always conjectural, never ultimate. In the *Theory of Harmony* (note 35), p. 8, Schoenberg wrote:

> I do not wish to quarrel with honest efforts to discover tentative laws of art. These efforts are necessary. They are necessary, above all, for the aspiring human mind. Our noblest impulse, the impulse to know and understand ..., makes it our duty to search. And even a false theory, if only it was found through genuine searching, is for that reason superior to the complacent certainty of those who reject it because they presume to know ... Since we do definitely know the phenomena [as facts] we might be more justified in giving the name, 'science' … , to our [direct] knowledge … of the phenomena, rather than to those conjectures that are intended to explain them.
>
> Yet these conjectures, too, have their justification: as experiments, as results of efforts to think, as mental gymnastics—perhaps sometimes even as preliminary steps to truth.
>
> If art theory … could be satisfied with the rewards afforded by honest searching, then one could not object to it. But it is more ambitious. It is not content to be merely the attempt to find laws; it professes to have *the eternal* laws. It observes a number of phenomena, classifies them according to some common characteristics, and then derives laws from them. That is of course correct procedure, because unfortunately there is hardly any other way. But now begins the error. For it is falsely concluded that these laws, since apparently correct with regard to the phenomena previously observed must then surely hold for all future phenomena as well.

artistic vision, because of what he conceives to be the irrational, relativistic nature of the composer's theory of art. Indeed, Popper's ideal republic, like that of Plato, would seem to be one in which the philosophers are kings, for Popper believes that correctness of interpretation is achieved not through the prophecy of intuition but through the prophecy of dialectic.

Of course, Popper is concerned almost exclusively with the logic of discovery. He states that he early abandoned investigations into the psychology of discovery (and, hence, into the sources of inspiration), since these investigations, he holds, fall 'into the traps of psychologism, idealism, positivism, phenomenalism, even solipsism'.[45] But Popper seems to have forgotten that whatever the sources of inspiration, Schoenberg still faced, as do all artists, the task of presenting his discoveries not only as musical compositions but also as writings on music. Consequently, Schoenberg had to reconcile the antithetical domains of Dionysus and Apollo, as have many others before him.[46] In fact, this reconciliation took place in Greece, for Dionysus

> had to be accorded a recognized and official position which would curb the wild irresponsibility that had marked his first entry into Greece and bring him as much as possible into line with the ... ancestral tradition and conventions of the Greek people. And who was better qualified to do this than Apollo, the 'national expositor' of

---

[45] Popper, *Unended Quest* (note 15), p. 76. See also K. R. Popper, *The Logic of Scientific Discovery* (New York, 1959) and *Objective Knowledge: An Evolutionary Approach* (Oxford, 1972). Certainly, Popper does not believe that man is wholly rational, for in *Conjectures and Refutations* (note 28), p. 357, he reveals an awareness 'not only of the power of emotions in human life, but also of their value'. Nonetheless, he admits he is inclined 'to protest against any exaggerations (arising largely from a vulgarisation of psychoanalysis) of the irrationality of man and of human society'.

[46] Schoenberg, *Style and Idea* (note 14), p. 414, wrote that the Dionysian-Apollonian antithesis—'the one, in intoxication, smashes the glasses which the other has produced in an intoxication of the imagination'—is not so simple: 'Things happen thus only ... in the imagination ... of a musicologist', although he supposed that intoxication, 'whether Dionysian or Apollonian, of an artist's fantasy increases the clarity of his vision'.

these traditions? They did therefore the best that they could do: they gave Dionysos a place at Delphi.[47]

[47] Guthrie, *The Greeks* (note 1), p. 201.

2. A CONJUNCTION OF CONCEPTS
MUSIC AND THE EMERGENCE OF PROBABILITY

Probability is the science of the laws of irregularities.
C. S. Peirce, *Collected Papers* ... (1931–58, 2. ¶769)

INTRODUCTION

Games of chance, including drawing of lots, are known to have existed at the beginning of civilisation. But historians of science claim that games did not facilitate the development either of combinatorial techniques or of concepts of randomness and probability. As contributory reasons for this, they cite the imperfection of ordinary dice, the belief in supernatural intervention and the comparative complexity of the rules of many games. By the middle of the seventeenth century, so the story goes, mathematicians such as Blaise Pascal and Pierre de Fermat became interested in the stochastic aspect of games of chance. These games provided them with precise examples of problems, the solution of which simultaneously led to the development of general principles and theory.

The traditional history of science approach fails to answer a particular question: why did the development of a systematic theory of random phenomena commence during the second half of the seventeenth century, when games of chance existed long before that time? In 1975, Ian Hacking attempted to answer this question by offering a different historical interpretation—that new concepts are generated from old by a process comparable to a mutation.[1]

---

[1] I. Hacking, *The Emergence of Probability: A Philosophical Study of Early Ideas about Probability, Induction and Statistical Inference* (Cambridge, 1975).

According to him, this interpretation accounts for the dual aspect of probability that, he claims, emerged in the time of Pascal. The first aspect, which is epistemological, is connected with the degree of belief warranted by evidence; the second aspect, which is statistical, is connected with the tendency displayed by some chance devices to produce stable frequencies.[2] In Hacking's opinion, probability in its modern dualist sense became possible only after the medieval concept of probability as testimony was conjoined with the notion of frequency, that is, with the acceptance that stable and law-like regularities are observable and worthy of observation as part of the technique of understanding our world. It is this 'conjunction' of concepts that resulted in what he calls 'a mutation'.[3]

Hacking also claims that from 1660 there was a new instrument of probability—the mathematical art of combinations and permutations. Although he follows traditional historians in associating the statistical or combinatorial aspect of probability with gambling problems, he devotes some attention to pre-seventeenth-century developments when, for example, he states:

> Roughly speaking, combinatorial problems remain thoroughly in league with the alchemical magic of signs until the sign itself is liberated from that background in the seventeenth century. The great alchemist Raymond Lulle ... is usually cited as the founder of the theory of combinations. He hoped to represent all the elements of the world by their true signs and then, by generating all possible combinations of signs, to produce true signs for all possible compounds in the universe. One would then know how to make any possible thing. The great combinatorial work follows this tradition.[4]

---

[2] Ibid., p. 1, where Hacking states: 'Neither of these aspects was self consciously and deliberately apprehended by any substantial body of thinkers before the time of Pascal.'

[3] The terms 'conjunction' and 'mutation' are part of the semantic field of Hacking's covert alchemical model. I admit to having had some fun with these terms, the former of which is retained herein but without its alchemical meaning. Since the latter term, in its Latin plural (*mutationes*), signifies combinations and permutations, I have used one or the other term or variants thereof.

[4] Hacking, *The Emergence* (note 1), pp. 49–50; see also J. N. Hillgarth, *Raimon Lull and Lullism in Fourteenth-Century France* (Oxford, 1971). Despite Lull's

Hacking, then, brings probability and its mathematical instrument into conjunction with a number of subjects previously ignored by historians—for example, alchemy and the alchemical magic of signs. Nevertheless, his thesis is no more successful than the traditional explanation, because both accounts value the past merely for its anticipations of, or steps towards, some future specified goal. Consequently, much of the data of ancient and medieval times are ignored, although without such data, no manifest account can be given of historical change. Indeed, Hacking is unable to provide such an account, for the process by which his so-called 'mutation' takes place remains occult.

In 1980 Daniel Garber and Sandy Zabell produced a detailed critical analysis of Hacking's thesis.[5] They showed convincingly that the supposedly modern concept of probability is not modern at all but was widespread before the seventeenth century, particularly in the literature on rhetoric.[6] For Garber and Zabell, therefore, the historical problem is reducible to two questions: 'why did the theory of games of chance, perhaps already known in some of its basics, suddenly attract the attention of mathematicians as an interesting mathematical problem, and why did it not receive that attention long before the

---

importance in the history of combinations, he was not the founder or inventor of that art. In the first place, he never formalised his system; in the second place, the practice of meditating on combinations of letters was, before Lull, a Jewish phenomenon, developed particularly in the Spanish Cabbala as the meditations on combinations of the sacred Hebrew alphabet. The process of formalisation seems to have taken place between the twelfth and the fourteenth centuries, when general mathematical rules were enunciated not only in connection with combinations of the Hebrew alphabet but also with other sorts of combinations—e.g., the number of possible conjunctions of celestial mobiles.

[5] D. Garber and S. Zabell, 'On the emergence of probability', *Archive for History of Exact Sciences* (1979-80), 21: 33–53.

[6] See also O. B. Sheynin, 'On the prehistory of the theory of probability', *Archive for History of Exact Sciences* (1974), 12: 97–141, who includes some Arabic, Chinese and Indian contributions; and N. L. Rabinovitch, *Probability and Statistical Inference in Ancient and Medieval Jewish Literature* (Toronto, 1973). These two studies provide abundant evidence that probabilistic reasoning is not the characteristic creation of recent centuries and which, by virtue of its modernity, has rapidly become the major mode of rational inquiry.

middle of the seventeenth century?'[7] Their answer to these questions proceeds on sociological and psychological grounds.

While Garber and Zabell are undoubtedly correct that probability in its dual aspect is much older than the seventeenth century, I cannot agree that the historical problem is solved by sociological or psychological explanations alone. I propose, therefore, to offer a different line of argument by focusing on cosmology as one of the keys to the emergence of mathematical probability and by utilising evidence from the philosophy, theory and practice of music. Because of music's traditional association with cosmology and with mathematics, it seems a better candidate for studying aspects of the prehistory of probability than does the traditional historians' gambling problems, Hacking's alchemical magic of signs or Garber and Zabell's sociological and psychological explanations.

In what follows, then, I provide a fourteenth-century instance of the supposedly modern dual aspect of probability, in which there is a conjunction of testimony and frequency. As I have shown elsewhere, the instance in question is a debate employing issues in the philosophy of music to examine the fundamentals of astrology and, in particular, the analysis of planetary aspects.[8] At a deeper level, however, the debate concerns order and disorder, that is to say, uniformity and variety, concord and discord, permanence and change. Because these issues have cosmological ramifications, a choice must be made between the positions taken in the debate. I therefore provide the background from which a liberally educated, fourteenth-century person could have determined the best decision. A central feature of that background is harmonic analysis—the application of theories of proportion to the analysis of macroscopic phenomena.[9] Since probability was restricted to the qualitative sciences of the microcosm, conditions were not favourable for bringing probability into conjunction with the mathematics of harmonic analysis. Hence,

---

[7] Garber and Zabell, 'On the emergence' (note 5), p. 48.

[8] J. C. Kassler, 'Music as a model in early science', *History of Science* (1982), 20: 103–139.

[9] Harmonic analysis has a history, as we shall see. For the modern mathematical version, see S. Bochner, *Harmonic Analysis and The Theory of Probability* (Berkeley, 1955).

in the debate the instrument of probability is logical, not mathematical.

By the seventeenth century there was increasing awareness of the usefulness of mathematics as an universal instrument for analysing not merely macrocosmic but also microcosmic phenomena. As a result, some writers supposed that the art of combinations (including permutations) could serve as the universal instrument of harmonic analysis. Hence, I consider some aspects of this art, particularly as it is found in music theory and applied in musical practice. Finally, I offer a sketch of a new historical thesis by arguing that in the seventeenth century the Cartesian mechanical philosophy brought about changed conditions that were more favourable to a conjunction of probability and mathematics. I outline how this change was effected and may be accounted for in a manifest way.

PROBABILITY AS TESTIMONY AND FREQUENCY

Before the seventeenth century the calculation of risks was made by casting lots. The methods involved ranged from determining the outcome of games to predicting the fortunes of individuals.[10] In the latter case, prognostications about events in the lifetime of individuals were part of divination or judicial astrology, which the church regarded as a magical rather than a mathematical science.[11] Nevertheless, practitioners of the magical science based their predictions on precise determinations of successively repeated conjunctions, oppositions and other planetary aspects. In order to

[10] For an account of one medieval method for determining by means of music theory the outcome of a game, see J. Hawkins, *A General History of the Science and Practice of Music* [1776] (2 vols., New York, 1963), vol. 1, pp. 51–52.

[11] The church tolerated astrology as theoretical and applied astronomy, regarding these as legitimate studies insofar as they consisted of sciences of the worlds that fall within the range of our sensible perception; but it proscribed astrology as divination. The basis for this proscription is to be found in the scholastic distinction of true and false mathematics, the latter of which was regarded as a magical science. See, e.g., R. Bacon, *The Opus majus* tr. by R. B. Burke (2 vols., New York, 1962), vol. 1, p. 261.

undermine the predictions of astrologers, therefore, it was necessary to render astrology fallacious at the very beginning of its operations.

This was a principal aim of the fourteenth-century scholastic, Nicole Oresme, who, in a number of writings, argued that there was a kind of basic numerical indeterminacy which even the best astronomical data cannot overcome. His most strikingly stated case against the astrologers occurs in a treatise on commensurability and incommensurability, in which he proposed three things: first, to consider what follows if all the celestial motions are mutually commensurable; second, to consider what follows if some are incommensurable; and third, to examine whether a categorical decision may be made about the one or the other hypothesis.[12]

Oresme's examination of the decision problem takes place in a dream, where the dreamer encounters Apollo and the Muses and Sciences. Apollo at once rebukes the dreamer with these words:

> Because you are ignorant of the ratios relating the things of this world, your occupation is an affliction of the spirit and an unending labor. You should understand that exactness transcends the human mind. But, since you inquire about sensible things, you ought to begin with the senses through which ... exact precision is undetectable. For if an imperceptible excess—even a part smaller than a thousandth—could destroy an equality and alter a ratio from rational to irrational, how will you be able to know a punctual [or exact] ratio of motions or celestial magnitudes?[13]

The dreamer accepts that 'it is not given to human powers to discover such things unless furnished with some principle assumed beforehand'; that 'the judgment of the senses cannot attain ... exact knowledge' and that the problem of celestial motions and magnitudes cannot be solved by a mathematical demonstration.[14] Nevertheless, he wishes to assuage his doubt and to know whether or not motions of mobiles are commensurable. So Apollo commands the Sciences of

---

[12] E. Grant, *Nicole Oresme and the Kinematics of Circular Motion: Tractatus de commensurabilitate vel incommensurabilitate motuum celi* ed. with an introduction, English translation, and commentary (Madison, 1971).

[13] Ibid., p. 285.

[14] Ibid., p. 287.

Arithmetic and Geometry to teach the dreamer what he asks by defending 'their cause with reasoned arguments, as if they were litigants in a law suit'.[15]

There follows a debate between Arithmetic and Geometry, the structure of which is schematised in *Figure 2.1*. Each Science puts her case separately. Arithmetic, who commences the debate, holds that all motions of mobiles are commensurable, since incommensurable magnitudes are irrational and unreasonable—the divine mind would never connect motions of mobiles 'in such a haphazard relationship, when, indeed, it ought to arrange them rationally and according to rule'.[16] Hence, Arithmetic takes the position of a realist, for she holds that abstract entities lie outside space and time.

In putting her argument, Arithmetic presents a set of possible outcomes or consequences if the hypothesis opposite to hers were adopted. This set of outcomes may be summarised according to three general points: first, that incommensurability destroys the beauty of the universe and detracts from the goodness of the gods; second, that geometry (magnitude) brings harm to the established harmonic order of number (multitude); and third, that ignorance is produced in those who deny that the velocities of mobiles are proportioned by numbers, since if this were true, no predictions of the future would be possible. According to Arithmetic, none of these consequences would occur if the hypothesis of commensurability were adopted. To support her claim, she provides 'witnesses', that is, the testimony of twenty-three wise men. By means of these witnesses, Arithmetic hopes to prove her point of view.

Next comes Geometry, who argues that only some—not all—motions of mobiles are incommensurable. In her rebuttal of Arithmetic, Geometry takes a nominalist stance and asserts that 'nothing prevents false propositions from being more probable than certain true ones'.[17] According to Geometry, Arithmetic's case is less probable than her own, and she will show this with 'stronger

---

[15] Ibid., p. 289.

[16] Ibid., p. 291.

[17] Ibid., p. 311.

UNIVERSE

A Product of Man's Cogitation

ARITHMETIC

Order:	Uniform and regular
Relation:	Determinate (rational numbers, rational ratios)
Future:	Can be predicted with certainty, because commensurable motions result in the same or similar motions and effects being repeated, the ages returning in a circle

GEOMETRY

Order:	Difform and regular
Relation:	Indeterminate (mixture of rational and irrational numbers and ratios)
Future:	Can be predicted only probably, because the mixture of commensurable and incommensurable magnitudes means that the same event would be repeated infrequently; rather, new and dissimilar configurations would emerge from previous ones and would always produce different effects, the ages proceeding endlessly in a straight line

Figure 2.1. The debate between Arithmetic and Geometry

arguments and far less special pleading'.[18] To do this, she discusses Arithmetic's three general consequences briefly and 'as they bear upon the inferences to be drawn' from her own proposals.[19]

Geometry argues that if her hypothesis were adopted, the beauty of the universe and the goodness of the gods would increase, since there would be a mixture of uniform and difform elements. Moreover, no harm would come to the harmonic order of multitude, since that order is included within magnitude. Finally, only gods know the future with certainty and exactitude; since man is not a god, he can attain only approximate knowledge of such future contingents as celestial conjunctions. Hence, any man who 'announces results that are free of noticeable error would seem to have determined [such future] things adequately'.[20]

To support her case, Geometry employs empirical vindication of different kinds, using inferences that depend on sampling. For the first set of consequences, she has recourse to the testimony of the senses by referring to law-like regularities—that is, varieties—we may perceive in nature and in art. For the second set of consequences, she points out that Arithmetic's witnesses offer mutually conflicting evidence, that their physical measurements are faulty and that their statements are incompatible with the phenomena observed thus far by astronomers. For the third and last set of consequences, Geometry refers to mathematical demonstrations that show, for example, there probably are more irrational than rational ratios of ratios.[21]

It is to be noted that the premisses of both disputants are contingent and, therefore, initially possible in virtue of their contingency. But in the case of Arithmetic and Geometry, the premisses are possible in different ways. Those of Arithmetic arise from the idea that the probability of a proposition can be determined by a consensus, an idea traceable to Aristotle. Although Aristotle is

---

[18] Ibid.

[19] Ibid.

[20] Ibid., p. 319.

[21] For this notion, which is peculiar to Oresme, see the first treatise in N. Oresme, *De proportionibus proportionum and Ad pauca respicientes* ed. with introductions, English translations, and critical notes by E. Grant (Madison, 1966).

credited with inventing the syllogism, he did not overlook the importance of non-deductive inference, for he distinguished between demonstrative and dialectical reasoning.[22] In demonstrative reasoning the premisses are true and primary or are such that our knowledge of them has originally come through premisses that are true and primary. The result of such reasoning, according to Aristotle, is scientific knowledge that is true and incontrovertible.

In dialectical reasoning, however, the result is not certain. This is so, because the premisses are opinions, and opinions admit of error. Hence, dialectical reasoning was considered inferior, as is clear from Apollo's remarks at the conclusion of the debate, where he cautions the dreamer:

> Do not seriously believe that there is a genuine disagreement between these most illustrious mothers of evident truth. For they amuse themselves and mock the stylistic mode of an inferior science. In conversing with them, we also jestingly adopt the manner of doubt in our judgments.[23]

In enumerating the sources of opinion used in dialectical reasoning, Aristotle identified one source as probability, namely, testimony—that which is accepted by most men or by the wisest among them. It is this type of probable premiss that is used by Arithmetic. The premisses of Geometry do not arise from a consensus of witnesses. Instead, they arise from the idea that the probability of a proposition can be determined from the frequency of elements in a set, regardless of whether these frequencies occur in nature, in art or in mathematics.

---

[22] See Aristotle, *Topica* (100a27, 100b21–23); see also E. H. Madden, 'Aristotle's treatment of probability and signs', *Philosophy of Science* (1957), 24: 167–172.

[23] Grant, *Nicole Oresme* (note 12), p. 323. Apollo's remarks are made in answer to several of the dreamer's queries that

> Since every truth seemed consonant with each side, in what sense are these disagreements the parents of truth? And what do they really contribute with their rhetorical persuasions or sophistical proofs, seeing how contemptuous they are of every other kind of argument usually employed solely in demonstration? Why do they adopt the method of more uncertain knowledge—a method to which they are unaccustomed?

There is, however, a third conception of probability, for Arithmetic and Geometry offer us a pair of contradictory contingent statements. Since we do not know which statement is true, each statement is equally possible because of this doubt.[24] And here, perhaps, is the main point of the debate, for Oresme is not concerned with knowledge of individual things in themselves but with rational grounds for acting in the absence of such knowledge. But just as Apollo is to judge which hypothesis is the more probable, the dreamer wakes and the dream comes to an end. Hence, we are free to make our own judgment; but in so doing, we must be aware of the consequences of our choice, for, as the debate shows, cosmological arguments will be seriously affected by the adoption of the one or the other hypothesis.[25]

THE DECISION PROBLEM

A fourteenth-century, liberally educated reader of the debate would have recognised at once that Arithmetic recapitulates the central tenets of Pythagorean world harmony, the dominant cosmology of the early Middle Ages. That cosmology assumes an unfailing regularity of motions of mobiles, so that, on the basis of this assumption, the aim of astronomy is determined by a desire 'to save the phenomena', that is, to explain away the irregularities of planetary motions by a procedure which restores their supposed uniform conditions. This procedure may be termed 'harmonic analysis'. Deriving from the

[24] In his treatise *De pauca respicientes*, Oresme distinguished two general uses of the term 'possible'; for which see Oresme, *De proportionibus proportionum* (note 21), p. 385.

[25] The hypotheses of Arithmetic and Geometry might be equally probable when argued in terms of reason and experience, but only Geometry so argues. Arithmetic relies on authority that is shown to consist of conflicting witnesses who offer faulty measurements and whose observations disagree with astronomical data. Hence, it seems likely that Oresme himself regarded Arithmetic's hypothesis as improbable or even impossible; but his construction of the decision problem as a dream allows for the exercise of free will—and avoids the imputation to Oresme of heterodox conceptions.

techniques of music theorists, two approaches to harmonic analysis may be identified.[26]

One approach was modelled on the techniques of the early Pythagorean music theorists, for whom the basic facts of music were derived from speculation and everything else stemmed from operations in arithmetic. Hence, their eternal and immutable verities were arithmetic as well as music, because they regarded music not only as an arithmetical science but also as the purest embodiment of unchanging law. In their calculations of the scale of music, the Pythagoreans employed ratios based only on the prime factors of 2 and 3, thereby restricting the consonant intervals in music to the octave, with the proportion 2:1, the fifth, with the proportion 3:2 and the fourth, with the proportion 4:3, or to compounds of these—for example, the octave plus fifth, with the proportion 3:1. This approach was founded not on mathematics or on experimental physics but on reverence for certain numbers in their roles in music and cosmology.

The other approach to harmonic analysis was modelled on the techniques of the Aristoxenian music theorists, so-called from the name of their founder, Aristoxenus, a pupil of Aristotle and a contemporary of Aristotle's successor, Theophrastus.[27] For the Aristoxenians, the basic facts of music derived from experience and experiment, and everything else stemmed from the functional relation of musical elements. To establish intervals in the scale of music, the Aristoxenians relied on geometrical operations, for they represented the whole tone as a line and then divided the line. By means of this spatial metaphor, equal intervals could be symbolised as equal distances or lines, and greater intervals could be made up of the sums of smaller intervals. This approach was founded on the attempts of Aristoxenus to construct a theory that corresponded with the facts of musical practice and to handle musical intervals without reference to numbers or ratios.

---

[26] For this kind of analysis, see Kassler, 'Music as a model' (note 8), and sources cited there.

[27] R. D. McKirahan, 'Aristotle's subordinate sciences', *The British Journal for the History of Science* (1978), 11: 197–220, p. 208, describes Aristoxenus's procedure as 'revolutionary'.

In the Pythagorean approach to harmonic analysis, intervals were defined through proportions that were computed by means of Pythagorean arithmetic. A principle of that arithmetic stated the impossibility of inserting geometric means between the terms of superparticular proportions—that is, proportions in which the terms differ by 1—for such a process would involve irrational numbers. It is this principle that provided the arithmetical basis for the Pythagorean belief that the whole tone, represented by the superparticular proportion 9:8, is not divisible into two or more equal parts.

About 1317, Marchetto of Padua proposed a system in which the whole tone is fractionally divisible, thereby abandoning the basis of Pythagorean harmonic theory. In so doing, he provided 'the initiative for the ensuing centuries' manifold experiments with fractional division of the whole tone', experiments that would lead eventually to equal temperament and modern harmony.[28] Along with these experiments there would be a change or extension in the definition of consonance, a change that would have profound implications for cosmology, because only the consonances of the octave, fourth and fifth supported the Pythagorean harmonic order.

Although that harmonic order lasted until the time of Johannes Kepler, it was severely challenged by sceptical empiricists of the thirteenth and fourteenth centuries. Such a challenge occurs in the debate constructed by Oresme, where Geometry's use of irrational numbers is subversive to the reign of Arithmetic and her supposed uniformity and regularity of celestial mobiles, for irrational numbers are not uniform. Moreover, on Geometry's assumption of variability, there can be no eternal, unchanging verities based on the prime factors of 2 and 3, for the observable world cannot be reduced solely to whole numbers or to comparisons of commensurable magnitudes.

Now, it is important to grasp some of the implications of the two different approaches to harmonic conceptions and harmonic analyses, for they articulate different value systems. The Pythagorean concern was with invariable, eternal, ideal forms, the model of which was a proportional tuning system that emphasised stasis. The Aristoxenian concern was with variability, the model for which was a

[28] J. W. Herlinger, 'Marchetto's division of the whole tone', *Journal of the American Musicological Society* (1981), 34: 193–216, p. 215.

scale that deviated from the 'pure' intervals of Pythagorean tuning, thereby allowing for dynamic, expanding and novel means of expression.[29] In Oresme's debate, Arithmetic represents Pythagorean harmony, for she attempts to preserve the traditional static order. Geometry, however, represents Aristoxenian harmony, for she attempts to embrace dynamic change.

The antithesis, then, is between order and disorder. But disorder does not signify chaos—it merely implies defective order. To think about a disordered state, we must have in mind an ideal of order from which it falls short. It is much easier to characterise disordered systems in terms of their deviations from this ideal than it is to define a perfectly disordered system on which some partial degree of order is to be imposed. The concept of disorder is primitive and intuitive: it belongs with statistical terms such as 'random', 'stochastic', 'unpredictable', which can only be defined within a specific context of what is already known or can be taken for granted.

In the writings of Oresme, arguments appear in which stochastic considerations are applied to an abstract mathematical notion, so that in his speculations we find a form close to modern mathematical probability. For example, in his treatise on ratios of ratios, Oresme applies the mathematics of combinations to probability.[30] But in the debate between Arithmetic and Geometry,

---

[29] Geometry's position also reflects the increasing individualism of the period, as exemplified by the composer, Guillaume de Machaut. See, e.g., S. J. M. Williams, 'Machaut's self-awareness as author and producer', *Machaut's World: Science and Art in the Fourteenth Century* ed. by M. P. Cosman and B. Chandler (New York, 1978), pp. 189–197. Machaut was a contributor to the *ars nova* movement, an expressionist movement in music that, through the use of chromaticism, posed a threat to traditional conceptions of the structure of music and, hence, of the structure of the world.

[30] Oresme, *De proportionibus proportionum* (note 21), pp. 40–42, 247–263, 303–309, 362–363, 385. See also Rabinovitch, *Probability* (note 6), pp. 145–151. The theory of combinations was elaborated by the fourteenth-century mathematician, Levi ben Gershon, whose work was known to Oresme, probably through the intermediary, Philippe de Vitry, one of the prominent musicians of the *ars nova* movement. See E. Werner, 'The mathematical foundation of Philippe de Vitry's *Ars nova*', *Journal of the American Musicological Society* (1956), 9: 128–132 and Kassler, 'Music as a model' (note 8). Until the early modern period, the composer's art was a trade secret, so that music analysis is the chief means for determining whether combinatorics was applied in the

his main procedure for dealing with the contingent is the dichotomous instrument of disputation. That instrument, which is logical, was favoured by medieval scholastics. From the seventeenth century, however, the preferred instrument was mathematical.

A NEW INSTRUMENT OF ANALYSIS

The art of combinations is intimately connected with probability, for without its help the enumeration of equally possible cases cannot be accomplished. But, as one historian of probability points out,

> Today, one conceives of the mathematical theory of probability in purely formal axiomatic terms as a branch of measure theory, and the interpretation which connects the mathematical theory to reality is entirely extra-mathematical. The assignment of probabilities to events is solely a philosophical undertaking [so that] ... for the mathematician it is irrelevant whether there is any empirical or other correlate to his probability functions.[31]

John Venn thought similarly about the elementary theory of his day, for he wrote:

> When Probability is ... divorced from direct reference to objects, as it substantially is by not being founded on experience, it simply resolves itself into the common algebraical or arithmetical doctrine of Permutations and Combinations.[32]

---

construction of musical compositions. Unfortunately, however, the connections between the medieval theory of combinations and the practice of music have not been systematically investigated, although it is possible that one such connection might be discovered in the work of Marchetto of Padua, who added a new term, 'permutation', to the theory of musical mutation; see Herlinger, 'Marchetto's division' (note 28), pp. 196–197. See also H. Kirchmeyer, 'On the historical constitution of rationalistic music: A study of the progress of musical combinatorics from Guido d'Arezzo to the present day [tr. C. Cardew]', *Die Reihe* (1968), 8: 11–24.

[31] Rabinovitch, *Probability* (note 6), p. 142.

[32] J. Venn, *The Logic of Chance* (3d edn., London, 1888), p. 87.

Nevertheless, in what is regarded as the landmark work on the purely mathematical theory of probability—Jacques Bernoulli's *The Art of Conjecturing* (1713), two sections are devoted not only to the theory but also to the application of combinations.[33]

Bernoulli introduces these sections with a hymn on the infinite variety appearing in the works of nature as well as in the actions of humans. According to him, this variety accounts for the greatest part of the beauty of the universe. Here, then, we have the same point of view as Geometry. But whereas Geometry supposes the cause of variety to be a mixture or blending of commensurable and incommensurable magnitudes, Bernoulli asserts that variety in the world stems solely from the composition, combination and arrangement of its parts. Therefore, he argues, the art of combinations is extremely useful, because it teaches us to enumerate, without omission, all possible categories which allow us to mix, arrange or put together several things. Most important of all, the art of combinations is not merely a branch of mathematics but has a relation 'to almost every species of useful knowledge that the mind of man can be employed upon'.[34]

Just how commonplace the art of combinations was to become from the early part of the seventeenth century may be instanced by reference to some combinatorial problems that were studied in music

---

[33] J. Bernoulli, *Ars conjectandi, opus posthum. Accedit. Tractatus de seriebus infinitis, et epistola gallice scripta de ludo pilae reticularis* [1713] (Bruxelles, 1968). The treatise is considered a landmark in the history of probability, because the calculus of games of chance is brought into conjunction with the previously independent theory of probability for the first time. See, e.g., I. Schneider, 'Probability (mathematical)', *Dictionary of the History of Science* ed. by W. F. Bynum, E. J. Brown and R. Porter (London, 1983), pp. 338–339; see also Hacking, *The Emergence* (note 1), p. 166. I have used the translation in F. Maseres, *The Doctrine of Permutations and Combinations, being an essential and fundamental Part of the Doctrine of Chances: as delivered by Mr. James Bernoulli ... and by the celebrated Dr. John Wallis*...(London, 1795). It is noteworthy that cosmology and music are treated in conjunction (as they are in the work of Wallis), for after calculating the number of combinations of conjunctions of planets, Bernoulli turns his attention to the twelve registers or rows of organ pipes.

[34] Ibid., pp. 35–36.

theory and experimented with in musical practice.[35] In music theory the most notable of these studies may be found in Marin Mersenne's writings published after 1635.[36] These contain long sections devoted to combinations (for example, of organ registrations) and permutations (for example, of melody and rhythm), an example of which is illustrated in *Figure 2.2.* In the practice of music the art of combinations was used to devise methods for composing music by chance, that is, by random association of ready-made musical elements.[37] One such method, constructed by a publisher with a keen view to commercial success, was popularly known as 'Mozart's dice game'.[38] Another method has the title: *A tabular System whereby the Art of composing Minuets is made so easy that any Person, without the least Knowledge of Musick, may compose ten Thousand, all different, and in the most pleasing Manner* (c.1763).

Referred to by a nineteenth-century writer as 'tuneful lotteries',[39] these methods proliferated from about 1750. But one hundred years previously Athanasius Kircher had described how to construct such a method; and in the 1670s Samuel Pepys,

---

[35] There is no entry for combinatorics in *The New Grove Dictionary of Music and Musicians* ed. S. Sadie (20 vols., London, 1980). For more recent developments, see P. Lansky and G. Perle, 'Set', ibid., vol. 17, pp. 197–199.

[36] See E. Knobloch, 'Marin Mersennes Beträge zur Kombinatorik', *Sudhoffs Archiv* (1974), 58: 356–379, and 'Musurgia universalis: Unknown combinatorial studies in the age of baroque absolutism', *History of Science* (1979), 17: 258–275.

[37] For bibliographic particulars, see L. G. Ratner, '*Ars combinatoria*: Chance and choice in eighteenth-century music', *Studies in Eighteenth-Century Music: A Tribute to Karl Geiringer on his Seventieth Birthday* ed. by H. C. Robbins Landon and R. E. Chapman (London, 1970), pp. 343–363, and S. A. Hedges, 'Dice music in the eighteenth century', *Music and Letters* (1978), 59: 180–187. For British methods, see J. C. Kassler, *The Science of Music in Britain, 1714–1830: A Catalogue of Writings, Lectures and Inventions* (2 vols., New York, 1979). For the statistician's point of view, see M. G. Kendall, 'The book of fate', *Biometrika* (1961), 48: 220–222.

[38] *Anleitung zum Componiren von Walzern so viele man will vermittlest zeier Würfel ohne etwas von der Musik oder Composition zu verstehen* (Berlin, 1793).

[39] See *Ackermann's Repository of Arts*, 2d series (1826), 7: 58–59; see also Kassler, *The Science of Music* (note 37), vol. 2, pp. 1123–1126.

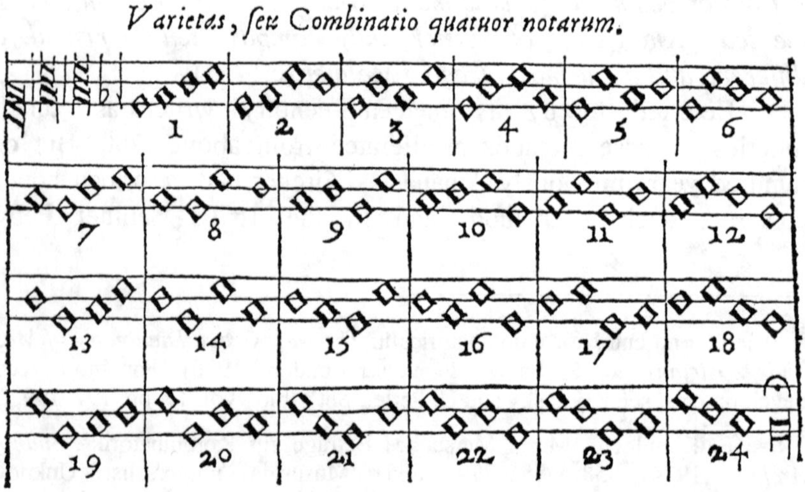

Figure 2.2. Varieties of Mi, Fa, Sol, La,
M. Mersenne, *Harmonicorum libri XII* (8th edn., Paris, 1648)

implementing Kircher's plan, produced his 'Musarithmica mirifica'.[40] Consisting of a miniature chest of 199 drawers in twelve tiers, each drawer has a small board inscribed with passages of music that may be variously combined to form melodies. The boards are similar to the 'bones' or calculating instruments described by John Napier in his *Rabdologia* (1617) and further developed by Kaspar Schott, whose *Organum mathematicum* (1668) included tables to aid the novice in composing music and creating melodies.[41] Most of the eighteenth- and nineteenth-century methods, however, consisted of passages of music printed on small cards or on sheets that could be cut into small cards as illustrated in *Figure 2.3*.

Methods for composing by chance were not confined to music, for Jonathan Swift, in his controversial fable, *Gulliver's Travels* (1726), described a logical calculus of words conceived as a combination of simple into complex concepts. In Swift's version ideas are like words: we all employ the same lexicon, and thinking is merely the concatenation of words according to the principles of a rudimentary probabilistic mechanics—the art of combinations.[42] But in *Gulliver's Travels* the 'thinking' is done by a kind of analytical engine or proto-computer, invented by one of the Laputans. It is described by Swift's anti-hero as follows:

> It was twenty Foot Square, placed in the middle of the Room. The Superficies was composed of several bits of Wood, about the bigness of a Dye, but some larger than others. They were all linked together by slender Wires. These bits of Wood were covered on every Square with Paper[s] pasted on them, and on these Papers were written all

---

[40] See R. T. Gunther, *Early Science in Cambridge* (Oxford, 1937), p. 96 and plate 107.

[41] M. R. Williams, 'From Napier to Lucas: The use of Napier's bones in calculating instruments', *Annals of the History of Computing* (1983), 5: 279–296.

[42] See J. C. Kassler, 'Man—A musical instrument: Models of the brain and mental functioning before the computer', *History of Science* (1984), 22: 59–92. In Swift's day there was a reassertion of natural law—that humankind is essentially the same, despite local, regional or national variations. This reassertion contributed to a tendency, during the Enlightenment, to take language as the model of all thought.

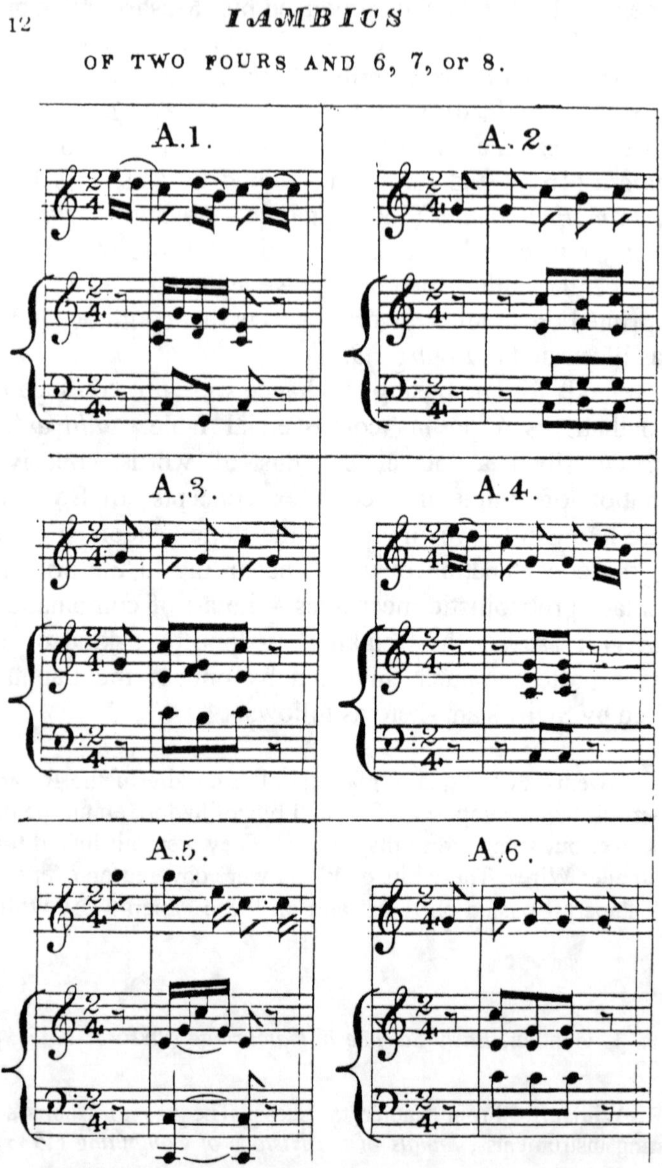

Figure 2.3. Method for composing music by chance, *Melographicon: A new musical Work, by which an interminable Number of Melodies may be produced...*(London, c.1825)

the Words of ... [the Laputan] Language in their several Moods, Tenses, and Declensions, but without any Order.[43]

Since the engine was not automatic, pupils of the Laputan inventor had each to crank one of the iron handles, of which there were forty fixed round the edges of the frame. We are told that upon giving the handles a sudden turn, 'the whole 'Disposition of the Words was entirely changed ... and at every turn the Engine was so contrived, that the Words shifted into new places, [as] the square bits of Wood moved upside down'.[44] Once stocked with words—which, in *Figure 2.4*, bear some similarity to the universal character invented in 1668 by John Wilkins and shown in *Figure 2.5*—the engine can generate a complete body of arts and sciences 'without the least Assistance from Genius or Study'.[45]

Not long after publication, in 1734, of the revised version of *Gulliver's Travels*, Francesco Geminiani constructed a composing 'engine' on the same principles as Swift's writing engine.[46] But Geminiani's engine is a dictionary of chords and chord progressions, which, in 1776, Charles Burney described pejoratively as 'a kind of mill, in which good music was to be ground with little trouble and no genius; as good sense and science by the Laputan machine in Gulliver's travels'.[47] Prior to its publication, however, the dictionary

---

[43] J. Swift, *Travels into several remote Nations of the World. In four Parts. By Lemuel Gulliver* (2 vols., London, 1726), vol. 2, p. 72.

[44] Ibid., p. 73.

[45] Ibid., p. 72. For the universal character, see J. Wilkins, *An Essay towards a real Character and a philosophical Language* (London, 1668). See also J. Wilkins, *Mercury: Or the secret and swift Messenger. Shewing, how a Man may with Privacy and Speed communicate his Thoughts to a Friend at any Distance* (London, 1641), where the author proposes that music may function not only as a universal character but also as a universal language.

[46] F. Geminiani, *Guida armonica, o dizionario armonico. Being a sure Guide to Harmony and Modulation in which are exhibited the various Combinations of Sounds, consonant, and dissonant, Progressions of Harmony, Ligatures, and Cadences, real and deceptive*...(London, c.1742). For details, see Kassler, *The Science of Music* (note 37), vol. 1, pp. 166–168.

[47] C. Burney, *A General History of Music, from the Earliest Ages to the Present Period* [1776-89] (2 vols., New York, 1957), vol. 1, p. 992.

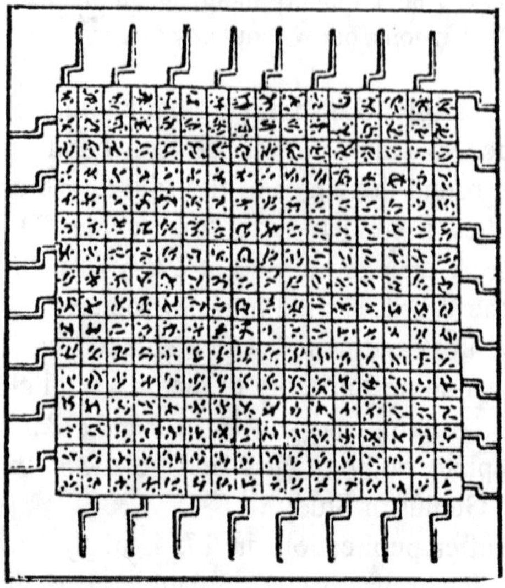

Figure 2.4. Method of writing by chance,
J. Swift, *Travels into several remote Nations of the World... by Lemuel Gulliver* (London, 1726)

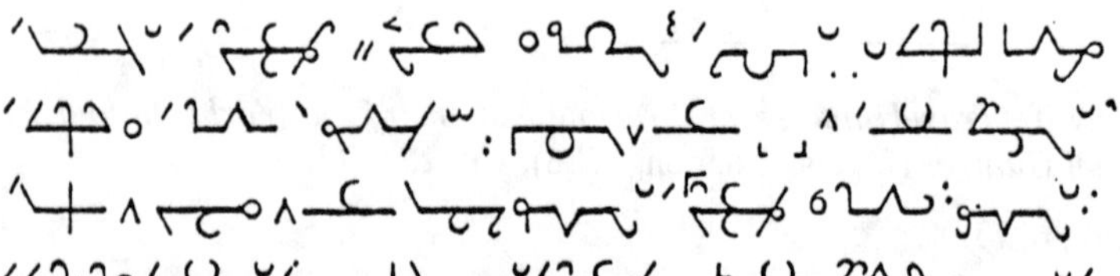

Figure 2.5. The universal character of John Wilkins,
R. Hooke, *Lectiones Cutlerianæ, or a Collection of Lectures...*(London, 1679)

had received a more favourable notice by Louis-Bertrand Castel, who detailed the combinatorial procedures upon which the dictionary had been constructed. According to him,

> It is ... plain that combinations alone make the riches of nature, and that art as well as its natural guide Geometry, may by means of it reach the infinite. In order therefore to embrace the infinite of musick, the whole consists in finding out all the elements of it, and all the immediate combinations of these elements.[48]

Since composition, for Castel, denotes 'harmony' and 'modulation'—that is, chords and their progressions, the elements of the former must be consonances, whereas the elements of the latter must be the 'passages or transitions and ties that link the consonances together, and constitute their coherency'.[49] In order to find out the most fundamental of these elements, Castel supposed that one could proceed by cutting up or dissecting entire pieces of music into small parts, some consisting of one measure, others of two half-measures. He then queried:

> Might not one next throw all these bits ad hoc & ab hac upon a paper, separating them all without any order or coherency, making however their natural succession by arithmetical figures, by letters and references, that would serve as a thread to put them in order, though it were for no other purpose but to procure to somebody the pleasure (a novice's pleasure indeed) of replacing them, and as it were of composing and creating something new?[50]

---

[48] L.-B. Castel, *An Extract out of the Journal des Scavans: Or Dissertation upon a Work wrote by Mr. Geminiani, intitled, The harmonick Guide* (London, 1741), p. 18. Castel is remembered chiefly for his 'Ocular Harpsichord', an instrument that produced visual, not musical 'harmony', thereby providing painters with a source of new colour combinations. Although a model was constructed toward the end of 1734, Castel spent most of his life trying to perfect the instrument, which was completed in July 1754. See Kassler, *The Science of Music* (note 37), vol. 1, pp. 166–168.

[49] Castel, *An Extract* (note 48), p. 18.

[50] Ibid., pp. 19–20.

There are two important points to note about Castel's suppositions. First, we have here a new conception of harmonic analysis, one that is different from the Pythagorean or Aristoxenian approaches to harmonic analysis described previously. The new method proceeds from the assumption that a piece of music may be decomposed into its material elements and recomposed or generated again from those elements.[51] This is the fundamental assumption of modern music analysis. Second, we have the method in which most of the devices for composing music by chance were constructed before as well as after Geminiani's dictionary. The fundamental differences between these devices are the given conditions in which the combinations are made.

CONCLUSION

The widespread interest in combinatorial problems from the early part of the seventeenth century coincides with the revival of materialistic atomism and the appearance of the mechanical philosophy of René Descartes. According to that philosophy, material elements in motion merely change their position or arrangement in time, thereby generating new configurations of material elements. In music theory, this combinatorial principle underlies the principle of chord inversion, whereby the notes constituting a triad—keynote, third and fifth—change their position and arrangement to form the first and second inversions of that triad, as shown in *Figure 2.6*.[52] The same principle could be applied to the treatment of larger musical structures such as

---

[51] For a similar assumption, see the definition of chemistry given by Ernst Georg Stahl and quoted in W. H. Brock, 'Analysis (chemical)', *Dictionary of the History of Science* ed. by W. F. Bynum, E. J. Brown and R. Porter (London, 1983), pp. 13–14.

[52] Roger North provided the first systematic formulation of the principle of chord inversion; see J. C. Kassler, 'Roger North', *The New Grove Dictionary of Music and Musicians* ed. by S. Sadie and J. Tyrrell (2d edn., 29 vols., London, 2000), vol. 18, pp. 53–56.

subjects, and it could be used also to account for the generation of new musical ideas in the mind.[53]

a b c

Figure 2.6. The principle of chord inversion, showing (a) root position *F–A–C*, (b) first inversion *A–C–f*, (c) second inversion *C–f–a*

In the mechanical philosophy God becomes a kind of Heraclitean inventor of a world that, like music, is compounded of *discordia concors*—a paradox, of course, since discord and concord are logical opposites. Indeed, mechanical philosophers of various persuasions seem to have been masters of paradox, including logical and mathematical problems that had not yielded to solution, curious machines and engines, illusions, games, tricks and, perhaps most importantly, scientific ideas contrary to general opinion. These paradoxes helped to contribute to an understanding of 'the ultimate mystery of metamorphosis, which can be practised upon nature by art, so that their collusion, their playing together may result in protean formations and deformations, in *serio ludere*'.[54]

---

[53] For a combinatorial treatment of the subject, see Chapter 8 *infra*; for an application of the combinatorial principle to ideas, see J. C. Kassler, *Inner Music: Hobbes, Hooke and North on Internal Character* (London, 1995), pp. 174–181.

[54] R. L. Colie, 'Some Paradoxes in the Language of Things', *Reason and Imagination: Studies in the History of Ideas 1600–1800* ed. by J. A. Mazzeo (New York, 1962), pp. 93–128, p. 108, who surveys a number of paradoxes, along with the joco-serious nature of the works in which they are contained.

It would seem highly probable, therefore, that the mechanical philosophy contributed to the seventeenth-century proliferation of combinatorial studies and to the conception of combinatorics as an universal instrument of analysis. Indeed, those studies and that conception predate the work of Pascal and Fermat, whose correspondence on the arithmetic of combinations is usually cited by traditional historians as contributing to the emergence of the purely mathematical theory of probability. For a conjunction of probability and mathematics to place, however, probability had to acquire an improved status. Its ancient connection was with the natural, not the exact sciences, and before the seventeenth century the natural sciences were considered inferior for two principal reasons. First, they were sciences of the microcosm, the sublunar world that was regarded as an imperfect reflection of the perfect mechanisms of the macrocosm. Second, certainty was precluded in studies of the sublunar world, because perfect knowledge was denied man inasmuch as earthly existence was imperfect.

The imperfection of earthly existence was linked with variability and change, as Arithmetic herself points out, when she states:

> if certain things near to us are sometimes proportioned incommensurably by virtue of a continuous variation, this happens—and I use the words of Aristotle—'because things in these lower regions have twisting [and confusing] motions as a result of their remoteness from proper divinity'—that is, from God, who would not permit such disorder near himself in the heavens.[55]

By admitting the possibility of variability in the macrocosm, Oresme drew from a long tradition in the natural sciences of the microcosm. These sciences were qualitative, whereas only the exact sciences were quantitative. As sciences of the macrocosm, where variability supposedly did not exist, the exact sciences never made variability a subject of speculation and investigation. Oresme's efforts notwithstanding, such an investigation had to wait until the

[55] Grant, *Nicole Oresme* (note 12), p. 317. Oresme's debate contains numerous paradoxes, many of which are expressions in a belief at variance with opinion in his day.

seventeenth century, when the mathematics of variability was to emerge.

In that century the Cartesian mechanical philosophy alters the old world picture by reversing the roles of macrocosm and microcosm. That philosophy teaches three things. First, the macrocosmic particles or ultimate material entities are insensible. Second, those particles and their various configurations cause everything that happens in the world. And third, since ultimate entities are insensible, we can understand them only by what we know on the microcosmic level. The terms 'macrocosm' and 'microcosm' now undergo change, the macrocosm becoming microscopic and the microcosm, macroscopic. Microscopic signifies the insensible, invisible, intangible bodies of the world; macroscopic denotes the sensible, visible, tangible bodies. Since, according to the mechanical philosophy, the latter are a key to understanding the former, we may conclude, for example, that such invisible elements as air particles move in simple harmonic motion only after we possess all the signifying analogies of bodies which we can see and put to the test of experiment, bodies such as pendulums, springs and musical strings.[56]

Starting with mechanics, therefore, the natural sciences, along with their methods of probable inference depending on fair sampling, gain in prominence. At the same time, the ground is laid for mathematics to become an universal instrument of analysis, for the

---

[56] See, e.g., M. Chan and J. C. Kassler (eds.), *Roger North's Cursory Notes of Musicke (c.1698–c.1703)*...(Kensington, N.S.W., 1986). To set forth his philosophy of music, North had recourse to a combination of hypothetico-deductive and inductive methods. The latter method was used in the natural history sections of the treatise, where North proceeded by making simple propositions, proving them by multiple instances, using two sampling techniques: statistical and rational. Hence, his demonstrations were presented as 'suppositions' imposed by hypothesis; in being confirmed by experiment or experiences, these then served to validate the hypothesis, promoting it to the rank of theory. As a lawyer North was experienced in evaluating testimonial, as well as circumstantial evidence, an experience gained, in part, from North's mentor, Matthew Hale, and from his brother, Francis North. For the importance of these two men in helping to introduce the law of evidence, see B. Shapiro, *Probability and Certainty in Seventeenth-Century England: A Study of the Relationships between Natural Science, Religion, History, Law, and Literature* (Princeton, 1983).

mechanical philosophy reconstructs the primary qualities of both microscopic and macroscopic bodies out of the mathematical concepts of number, motion and extension. The universal instrument of analysis is not limited to the art of combinations but includes other branches of mathematics that are developed from the seventeenth century. These conditions, I suggest, proved favourable to a conjunction of probability and mathematics and are at the root of sociological or psychological explanations of the emergence of the purely mathematical theory of probability.

3. MAKING INFERENCES
MUSIC AND THE MECHANISATION OF LOGIC

Thought is a thread of melody running through the succession of our sensations.
C. S. Peirce, *Collected Papers ...* (1931–58, 5. ¶395)

Historians of logic presume that the development of conceptions of inference began with the work of George Boole, so that they tend to neglect earlier accounts of the nature of inference.[1] In the seventeenth century, for example, Descartes' account involved a theory that inference must be grasped in an *intuitus*; a conception of humans as machines; and a belief that God freely creates eternal truths.[2] This account implies three things. First, logic is unnecessary as a formal discipline, because the human mind already has a kind of innate logical power that is simple and primitive (i.e., not further analysable). Second, in order to give a reasoned response, the human machine will be obliged to perform the reasoning itself. And third, if the principles are embodied in the machine, then it is *the maker* of the machine who does the reasoning for it.

If we transfer these implications from a natural to an artificial machine, we have the broad outlines of a project, undertaken by William Stanley Jevons (1835–1882), to mechanise the operations of logical inference. Son of an iron merchant, Jevons was brought up as a Unitarian and took science courses at University College, London, interrupting his education at the age of eighteen to take the job of

---

[1] G. Boole, *The Mathematical Analysis of Logic* (Cambridge, 1847)

[2] S. Gaukroger, *Cartesian Logic: An Essay on Descartes' Conception of Inference* (Oxford, 1989).

assayer at the newly established mint in Sydney.[3] In 1859 he left Australia to complete his education, graduating M.A. and accepting the post of tutor at Owens College (afterwards the University of Manchester, now John Rylands University). Subsequently, he became professor of logic and mental and moral philosophy, as well as professor of political economy. In 1876 he returned to University College as professor of political economy, resigning from the chair in 1880 because of failing health. Two years later, at the age of forty-seven, he drowned while swimming alone. Only recently has it been recognised that Jevons was 'one of the great Victorian polymaths', since he wrote on a range of subjects that, in modern terms, included not only political economy and sociology but also philosophy of science, mathematics, logic, and natural as well as biomedical science.[4]

Although a number of commentators have remarked on Jevons's love of music,[5] this aspect of his life has received little serious attention, perhaps because none of his publications was devoted *solely* to music. This situation did not change, even after a selection of Jevons's papers and correspondence was published between 1972 and 1981. Yet, in all seven volumes there are numerous references to music, musicians, musical instruments, music theory and musical inventions.[6] The subject of music occurs with particular frequency in the letters written from Australia to his sister, Henrietta. In one letter, for example, he reported his purchase of an harmonium, a metronome and quantities of music, to which he also added 'a superb and most convenient music-holder of my own design and manufacture'.[7] And in another letter he asserted that

---

[3] W. T. Gridgeman, 'William Stanley Jevons', *Dictionary of Scientific Biography* ed. by C. C. Gillispie (16 vols., New York, 1970-80), vol. 7, pp. 103–107.

[4] T. Inoue and M. V. White, 'Bibliography of published works by W. S. Jevons', *Journal of the History of Economic Thought* (1993), 15: 1–24.

[5] See, e.g., M. Schabas, *A World ruled by Number: William Stanley Jevons and the Rise of Mathematical Economics* (Princeton, 1990), pp. 12, 15, 62, 67.

[6] R. D. Collison Black and R. Könekamp (eds.), *Papers and Correspondence of William Stanley Jevons* (7 vols., London, 1972–81).

[7] Ibid., vol. 2, p. 309.

Music is always to me the same, a condition of my existence, a part of me. I believe I could live a life of Music.[8]

The editors remark that during his residence in Australia, Jevons 'wrote a book about music, and invented a new system of notation'; that he sent the manuscript 'chapter by chapter' to Henrietta; and that 'no trace of the work remains'.[9] Fortunately, Michael V. White has identified the 'book about music' as 'The art and science of music', a manuscript now located in the Jevons archive of John Rylands University Library (MS JA6/47/1–17).[10] Unfortunately, however, the new system of musical notation makes no part of that manuscript and has yet to be found.[11] The editors omitted the letters, or portions of the letters, mentioning that system, although they included one letter in which Jevons remarked: 'How stupidly contrived the *notation* of music is, is it not. I should improve it I think if I had much to do with it'.[12] This is the only hint relating to Jevons's invention. Another hint, however, may be discovered in Jevons's project to mechanise the operations of logical inference, to which I now turn.

Against the purely receptive, recording and classificatory activities of the human mind, Jevons tended to rehabilitate the constructive and anticipatory activities—hence his concern to devise

---

[8] Ibid., vol. 2, p. 319.

[9] Ibid., vol. 1, p. 32.

[10] Personal communication. Schabas, *A World ruled by Number* (note 5), p. 144 n.9, points out that there are several essays on music in the Jevons archive, including one that explores mathematical applications.

[11] Although some loose sheets in the Jevons archive relate to notation, it is difficult to determine in all cases whether they relate to musical or logical notation. At least one sheet, however, bears some similarity to the method of notation used for indicating the placement of pins on the barrel of a mechanical musical instrument, for which see P. Engramelle, *La tonotechnie ou l'art de noter les cylinders...*(Paris, 1775).

[12] Black and Könekamp, *Papers and Correspondence* (note 3), vol. 2, p. 227 (3 May 1856). For some earlier attempts at reforming musical notation, see J. C. Kassler, *The Science of Music in Britain, 1714–1830: A Catalogue of Writings, Lectures and Inventions* (2 vols., New York, 1979).

a method for drawing inferences rather than for testing formulae.[13] In his project to construct a logical machine, therefore, he sought to embody his own method, which he called 'the indirect method of inference'.[14] He supposed that since the parts of the machine would 'embody the conditions of correct thinking', the machine itself would become 'the embodiment of a true symbolic method or Calculus'.[15]

From the time of Descartes there have been a number of attempts to construct machines that perform computations, a prime example being the work of Jevons's contemporary, Charles Babbage. But, as Jevons himself pointed out, two schemes only might be considered as predecessors of his own project to mechanise the operations of logical inference.[16] One was the failed scheme of a mechanical syllogistic devised by the mystic, Raimon Lull; the other was the machine spun from the fertile imagination of Jonathan Swift and described in his masterpiece, *Gulliver's Travels*.[17] Apparently, Jevons regarded the latter machine as closest to his own project. But Swift's machine was stocked with a lexicon of words. Hence, his idea could not have been realised (because of the size of the lexicon) until an advance in the science of logic enabled production of a truly mechanical logic. Jevons credited this advance to Boole, whose 'quasi-mathematical' (algebraic) system of symbolic logic was

---

[13] See F. Enriques, *The Historic Development of Logic: The Principles and Structure of Science in the Conception of Mathematical Thinkers* tr. by J. Rosenthal (New York, 1968), pp. 187, 210–212; see also I. M. Bocheński, *A History of Formal Logic* tr. and ed. by I. Thomas (2d edn., New York, 1970), pp. 266–267 *et passim.*

[14] W. S. Jevons, 'On the mechanical performance of logical inference', *Philosophical Transactions of the Royal Society* (1870), 160: 497–518 and plates 32–34, pp. 500, 509. An abstract of this article was published in *Proceedings of the Royal Society of London* (1870), 17: 166–169; and the whole was reprinted in W. S. Jevons, *Pure Logic and Other Minor Works* ed. by R. Adamson and H. A. Jevons...(London, 1890), pp. 139—172. Unless otherwise indicated, my quotations are from the original article.

[15] Ibid., p. 517.

[16] Ibid., p. 498.

[17] For Lull's 'machine', see Bocheński, *A History of Formal Logic* (note 13), pp. 272–273; see also F. Yates, *The Art of Memory* (Harmondsworth, 1978), pp. 175–196 *et passim*. For Swift's machine, see Chapter 2 *supra*.

modified in the 1860s by Jevons himself as his indirect method of inference.[18]

Supposing that a complete solution of any logical problem may be effected 'by forming the complete list of combination, in which the terms of the problem can manifest themselves', Jevons devised an alphabetic notation to code the terms of the problem, and he found all possible combinations of, for example, the two syllogistic premises: All A is B, and No B is C.[19] By means of this 'abecedarium' or 'logical alphabet', he reduced the process of indirect inference to 'the repetition of a few uniform operations of classification, selection, and elimination of contradictories', whereby logical deduction becomes 'a matter of routine',[20] so that the abecedarium merely required the addition of keys, levers and pulleys to become a logical machine.

In 1865, therefore, after making a simple sorting machine or 'logical abacus' adapted to the lecture room, Jevons contemplated designing a more complicated machine 'adapted to extensive problems and arguments'.[21] In 1867 he engaged a clockmaker to construct the machine, although the design had yet to be finalised.[22] By the middle of 1868 Jevons decided that the machine would resemble either 'a large accordion, or a very small piano ... [of] twenty-one keys, exactly like white piano keys'.[23] Finally, in 1869, the machine was built on the latter design to resemble 'a very small upright piano or organ', as illustrated in *Figure 3.1*.

---

[18] G. Boole, *An Investigation of the Laws of Thought, on which are founded the Mathematical Theories of Logic and Probabilities* (London, 1854); but see A. Church, 'A bibliography of symbolic logic', *Journal of Symbolic Logic* (1936), 1: 121–218, (1938), 3: 167–212. See also I. Grattan-Guinness, 'The correspondence between George Boole and Stanley Jevons, 1863–1864', *History and Philosophy of Logic* (1991), 12: 15–35.

[19] Jevons, 'On the mechanical performance', abstract (note 14), p. 168.

[20] Ibid., p. 151.

[21] W. S. Jevons, *Letters and Journal*...ed. by his wife [Harriet Ann Jevons] (London, 1886), p. 213.

[22] Ibid., pp. 236, 238.

[23] Ibid., p. 241.

Figure 3.1. The logical machine
(the predicate-key C is depressed)
W. S. Jevons, *The Principles of Science...*(2 vols., London, 1874)

Of the twenty-one keys, sixteen letter keys represented

> the terms or letters A, a, B, b, C, c, D, d … employed in our logical notation. When letters occur on the left-hand side of a proposition, formerly called the subject, each is represented by a key on the left-hand half of the keyboard; but when they occur on the right-hand side, or as it used to be called the predicate of the proposition, the letter-keys on the right-hand side of the keyboard are the proper representatives.[24]

The five remaining operation keys stood for the copula (pressed when the verb, 'is', or its symbolisation, '=', is met); the final stop or output (pressed when a proposition is completed); the finis (pressed at the termination of an argument or to restore the machine to its initial condition as a *tabula rasa*); and two disjunctive conjunction keys (pressed when the conjunction, 'or', or its symbolisation, '.|.', is met). To work the machine, an operator merely had to '(1) *Read each proposition as it stands, and play the corresponding keys:* (2) *Convert the proposition and read and play the keys again in the transposed order of the terms*'.[25]

From this brief summary, it is possible to identify Jevons's indebtedness to the technology and semantic field of music. To begin with, there is the overtly stated relationship between his logical machine and a 'piano' or 'organ' (that is, harmonium). During his residence in Australia, Jevons bought a harmonium,[26] which is a free-reed keyboard instrument that uses wind under pressure supplied by compression bellows. Although the external shape of Jevons's logical machine is similar, he chose to call it a 'piano'. Either term, however, would have been appropriate, since his machine excludes the parts necessary for the production of sound (in the case of the piano, a sound board and strings under tension). But he retained the parts that a piano and harmonium share in common, namely, rods and levers.[27]

---

[24] W. S. Jevons, *The Principles of Science: A Treatise on Logic and Scientific Method* [1874] with a new introduction by E. Nagel (New York, 1958), p. 108.

[25] Ibid., pp. 111–112.

[26] See above (note 7).

[27] Jevons, 'On the mechanical performance' (note 14), pp. 508–509:

Jevons used the latter parts to couple the action of the keyboard with the logical abacus—the individual symbols on the face of the logical machine (see *Figure 3.1*).

The logical abacus functions like a musical score on the 'desk' of a piano or harmonium, since symbols on the abacus as well as on the musical score make visible a sequence of operations. The difference between the two is that the symbols on the former are analogues for a mechanical cycle of operations within the machine itself, whereas the symbols on the latter are analogues for the sequence of a composer's thought. In both logical machine and musical instrument, however, a performer must play the sequences by pressing down individual keys on a keyboard. As pointed out previously, Jevons's keyboard has two types of keys: operation and letter keys, as illustrated in *Figure 3.2*.

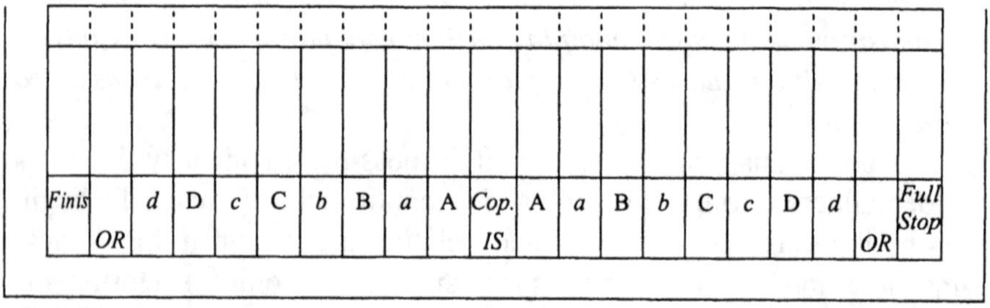

Scale 1/4

Figure 3.2. The keyboard of the logical machine

The operation keys are five in number—the copula in the centre; the finis and full stop on the extreme right and left; and the two keys next to the finis and full stop that represent the inclusive 'or'

The task of constructing a mechanical logic is thus reduced to that of classifying a series of wooden rods representing the conceivable combinations of the abecedarium into certain definite groups distinguished by their positions, and providing such mechanical arrangements, that wherever a letter term occurs in the subject or predicate of a proposition, or a conjunction, copula or stop intervenes, the pressure of a corresponding lever or key shall execute systematically the required movements of the combinations.

symbolised as '.|.'. This leaves the sixteen letter keys, which are arranged in two octaves, though restricted to the white keys of a piano or harmonium. In traditional musical parlance, these white keys may be named: ABCDEFG (the sequence is then repeated). Jevons, however, reduced the usual letters to the first four only: AaBbCcDd. And instead of repeating the same sequence of letters in each octave, the eight letter keys on the right are a mirror image of the eight letter keys on the left (just as the hands of a performer are mirror images).[28]

Finally, there is the connection between music and the conception of logic that Jevons embodied in his machine. He believed he had discovered a new kind of logic, which he called 'combinational logic' and which we now call 'the truth-table method'. This discovery was based on Jevons's assumption that the theory of combinations is the foundation of empirical science and mathematics; that nature also has a combinatorial character; and that there is 'an isomorphism between the assumed combinatorial character of the universe and the Logical Alphabet'.[29] It is noteworthy, therefore, that music, as an art of composition, has sometimes been referred to as 'Musica *Combinatoria*'.[30] Although the mathematical terms 'set' and 'combinatoriality' are common in twentieth-century music theory, the application of combinatorics and combinatorial probability to musical practice is much older. Witness, for example, the art of bell ringing called 'change ringing', first developed in England in the seventeenth century.[31] And witness the later eighteenth- and nineteenth-century methods for composing music by random association of ready-made elements, whereby a person with no training in music could still generate 'tuneful lotteries'.[32]

---

[28] Here is the one hint that may relate to Jevons's reform of musical notation.

[29] W. Mays and D. P. Henry, 'Jevons and logic', *Mind* (1953), 62: 484–505, pp. 489–490.

[30] J. Grassineau, *A Musical Dictionary; being a Collection of Terms and Characters, as well Ancient as Modern*...(London, 1740), p. 154.

[31] See W. G. Wilson, 'Change ringing', *The New Grove Dictionary of Music and Musicians* ed. by S. Sadie (20 vols., London, 1980), vol. 4, pp. 129–134.

[32] See Chapter 2 *supra*.

In 1874 Jevons alluded to this tradition when considering melodic combinations and permutations restricted to a single octave (the result is more than 40,000 permutations) and to a peal of twenty-four bells (they may ring continuously 'from the so-called beginning of the world to the present day').[33] Only a year earlier, however, Jevons's older contemporary, John Stuart Mill, wrote that he was

> seriously tormented by the thought of the exhaustibility of musical combinations. The octave consists only of five tones and two semitones, which can be put together in only a limited number of ways, of which but a small proportion are beautiful: most of these, it seemed to me, must have been already discovered, and there could not be room for a long succession of Mozarts and Webers, to strike out, as these had done, entirely new and surpassingly rich veins of musical beauty.[34]

Jevons (who disliked Mill's work on logic[35]) pointed out, first, that Mill had 'not bestowed sufficient study on the subject of permutations' and, second, that the end of 'novelty in art might well be dreaded, did we not find that nature at least has placed no attainable limit, and that the deficiency will lie in our inventive faculties'.[36]

Jevons's own inventiveness resulted in a machine that has been described as 'the first of its kind',[37] and as having more than antiquarian interest, because, operating, as it does, on simple mechanical methods, his machine 'illustrates in a clear and direct way the general principles upon which logical machines, electrical or otherwise, are based'.[38] But Jevons's logical machine was not *self-acting:* it required an operator to 'play' the keys. To become self-acting, a sequence control mechanism must supply the place of an

---

[33] Jevons, *The Principles of Science* (note 23), p. 191.

[34] J. S. Mill, *The Autobiography* (London, 1873), p. 145.

[35] See, e.g., Jevons, *Pure Logic* (note 14), pp. 175–299.

[36] Jevons, *The Principles of Science* (note 23), p. 191.

[37] M. Gardner, *Logic Machines and Diagrams* (New York, 1958), pp. 91–103, p. 100.

[38] Mays and Henry, 'Jevons and logic' (note 27), p. 493.

operator so that the machine can perform complex actions consisting of a sequence of perhaps quite simple basic operations. According to one historian,

> The earliest such mechanism was the rotating pegged cylinder still seen in music boxes today. Such cylinders are believed to date from the fourteenth century, when they were used to control the movements of model figures decorating large church clocks ... Later they were used to control pipe organs, and for example in the late sixteenth century [Salomon] De Caus used such a mechanism to control both the playing of an organ and several figures.[39]

Even as late as 1835 Babbage attempted to control the sequencing of the various component mechanisms for one of his engines (the analytical) by recourse to 'barrels' of the sort used in musical automata.[40]

Barrels, however, were regarded as an integral part of the machine they were controlling rather than as the means by which a general-purpose machine was set up to do a specific series of actions. As is well known, the idea of controlling a machine by sequencing information held on some separate medium seems to have arisen first in the silk-weaving industry.[41] By the mid- and late-nineteenth century, however, such schemes were used in automatic musical instruments. At first, the schemes consisted of interchangeable pegged cylinders and perforated discs. Later on, from the 1890s, the sequencing information was held on rolls of tape, as in early versions of the pianola or player piano.[42] In 1915 Leonardo Torres y Quevedo

---

[39] B. Randell (ed.), *The Origins of Digital Computers: Selected Papers* (2d edn., Berlin, 1975), p. 4, who refers to the automata of S. De Caus, *Les raisons des forces mouvantes avec diverses machines tant utiles que plaisantes* (Paris, 1615). For the music box, see, e.g., A. W. J. G. Ord-Hume, *Clockwork Music*...(London, 1973), p. 63.

[40] Randell (ed.), *The Origins of Digital Computers* (note 37), pp. 7-8.

[41] Ibid.

[42] Perforated paper sheets for producing music were first patented in France in 1842, according to H. Ellingham, *How to Use a Player-Piano* (London, 1922), p. 21; who describes, pp. 153–167, the way in which music rolls are made. See also A. W. J. G. Ord-Hume, *Player-Piano: The History of the Mechanical Piano and How to Repair it* (London, 1970).

proposed a means for constructing a variable automaton (chess player), based on that very technology, when he observed: 'It can be seen that the role of the human calculator who uses the machine can be reduced to pressing from time to time on certain keys, and that this may be automated by a process similar to that which has been used for operating a mechanical piano.'[43]

[43] Quoted in Randell (ed.), *The Origins of Digital Computers* (note 37), p. 95.

4. THE OVUM AS INITIAL DATUM
HARVEY, STARTING POINTS AND CREATIVITY

I have a young conception in my braine.
W. Shakespeare, *Troilus and Cresida* (1606, I.iii.312)

INTRODUCTION

In her marvellous study of Renaissance paradox, Rosalie Colie pointed out that seventeenth-century texts are preoccupied with creation and the 'beginning' of things.[1] But the beginning of things inevitably presents a puzzle: how do 'things', a whole set of ideas that can only with the greatest difficulty be thought away, 'begin'?[2] Is the starting point God who creates something out of nothing; is it pre-existing matter 'cogitating' or shaking out something; is it a vision, intuition or synthetic a priori coming into embodiment; or is it a freely chosen convention guided by experimental facts? The last answer comes close to our concept of creativity as novelty, the making of new things. But this definition dates from the nineteenth century, when creativity became the exclusive attribute of the artist and only later, in the twentieth century, was extended to other fields of human production.[3] Yet, this new concept was not itself created *ex*

---

[1] R. L. Colie, *Paradoxia epidemica: The Renaissance Tradition of Paradox* (Princeton, 1966), p. 61.

[2] For Thomas Hobbes's answer to this question, see T. Hobbes, 'Elements of philosophy. The first section, concerning body', *The English Works of Thomas Hobbes of Malmesbury* ed. by W. Molesworth (11 vols., Scientia Aalen, 1962), vol. 1, p. 92.

[3] See W. Tatarkiewicz, *A History of Six Ideas: An Essay in Aesthetics* (The Hague, 1980), pp. 92–94, 244–265.

nihilo, since, prior to the nineteenth century, the physician, William Harvey (1578–1657), conceived creativity as novelty, when, in his studies of the making of a new life, Harvey sought to make a new science.[4]

Harvey's theory is preserved in a book that brings together his embryological observations and experiments conducted over a period of more than twenty years. Published in 1651 as *Exercitationes de generatione animalium*, the book treats many problems at the frontiers of research and, hence, at the limits of discourse.[5] In assaying these problems Harvey relied on a number of different strategies, including that of informal disputation. This technique, which was taught in medical schools for examining a fact or theory in an attempt to ascertain the truth of a proposition made concerning it, required reference to the particular fact or theory under disputation, either by means of quotation, paraphrase or summary. It is noteworthy, therefore, that at the very outset of his book, Harvey summarised previous embryological opinion as follows:

> All Physicians, following Galen, teach that out of the seed of male and female mingled in coition, the offspring resembles either this or that parent according to the predominant power of the one seed or of the other, and that either a male or a female is begotten. And sometimes they pronounce the male's seed to be the efficient cause and the female's to serve as the material, and sometimes again the very opposite.
>
> But Aristotle (Nature's most diligent searcher) affirms that both the male and the female are the first principles of generation, that the female contributes the matter, the male the form, and that forthwith

[4] According to Hobbes, 'Elements of philosophy...concerning body' (note 2), vol. 1, p. viii, 'the science of *man's body* ... was first discovered with admirable sagacity by our countryman, Dr. Harvey, ... in his books of the *Motion of the Blood*, and of the *Generation of Living Creatures;* who is the only man I know, that ... hath established a new doctrine in his life-time'.

[5] All citations are to W. Harvey, *Disputations touching the Generation of Animals* [1651] tr. by G. Whitteridge (Oxford, 1981). For details of its composition and structure, see R. G. Frank Jr., *Harvey and the Oxford Physiologists: Scientific Ideas and Social Interaction* (Berkeley, 1980), pp. 34–38. Harvey's term, 'generation', has wider import than the term, 'embryology', which was introduced by Charles Darwin to refer to the study of the developmental process.

> after coition there is formed in the womb out of the menstrual blood the vital principle and the first particle of the future foetus, namely the heart in all sanguineous creatures.[6]

At the conclusion of this summary, Harvey remarked: 'That these are false and rash assertions will soon appear, and they will instantly vanish like phantoms of the night when the light of anatomical dissection dawns upon them'.[7]

Here, then, is the first hint that Harvey will present embryological ideas contrary to received opinion. But this is not the only type of paradox in Harvey's book, which includes such classical problems as beginnings and endings, being and becoming, time and eternity, the one and the many, as well as the Christian paradox of creation *ex nihilo*. These paradoxes arise, in part, because Harvey treated generation as an expression of the eternal nature of cyclical processes in the cosmos—for example, he wrote that just as 'the rising and the setting sun by its endless revolutions fills up the measure of time, so the transitory existence of mortal creatures is perpetuated by being forever reintegrated into the same species while individuals perish'.[8]

In addition, however, there are epistemological problems of self-reference, where Harvey draws attention to his own techniques, and self-reflection, where he reaches a terminus of thought and turns back to examine the process of thinking by which he arrived at an impasse. There are logical problems, including those called 'impossibles', as in Harvey's attempts to harmonise radical contradictions between his own opinions and those of Aristotle and the Galenic physicians. And there are a number of rhetorical problems. For example, Harvey's book might seem to cheat the reader's expectation, because it has no formal ending—it simply stops with an inconclusive chapter and then adds three treatises, the last of which is titled, 'Of conception'. Yet, these treatises are so constructed as to stimulate further thought and understanding in the

---

[6] Ibid., p. 8.

[7] Ibid.

[8] Ibid., p. 85; for Harvey's comment on the paradox of creation *ex nihilo*, see pp. 188–189.

reader and, hence, to awaken that reader to the importance of means—explicit observations; experiments by dissections, especially vivisections, utilising techniques such as ligature, inflation, injection and quantification; comparisons of the findings;[9] and, upon this foundation, the formation of physiological concepts.[10] Last, but not least among the rhetorical problems, there is Harvey's praise of the egg as both seed (beginning) and fruit (end), though, conventionally, the egg was a thing considered low.[11]

How are we to explain this veritable cornucopia of paradox, where one paradox seems to lead to the making of another? Colie provides a possible answer in her statement that even though paradox is always 'about' being, 'it is not fully ontological, since by drawing attention to its own form and technique it demands a "wondrer," a

---

[9] For details of Harvey's method, see Frank, *Harvey and the Oxford Physiologists* (note 5), pp. 16–18.

[10] According to Aristotle's interpretation of concept formation, the active intellect abstracts forms from the images, which are actual concepts when received in the passive intellect. In Harvey's day, there were three other interpretations of concept formation: Platonic, Epicurean and Stoic. On a Platonic interpretation, concepts are inborn ideas bearing testimony to the reminiscence of mankind's divine origin. René Descartes revived this interpretation, when he argued that concepts have nothing to do with the senses or the imagination; rather, they are purely intellectual and originate in a vision or intuition—the clear and distinct idea. On the Epicurean interpretation, concepts originate in sensation and are better labelled as percepts, whereas on a Stoic interpretation, concepts originate from a process of mental construction as principles expressive of the uniformity of nature. In the objections to Descartes' *Meditations* (1641), Pierre Gassendi took the Epicurean position and Thomas Hobbes, the Stoic one. For Gassendi, all ideas have their origin in sensation; imagination and conception are one and the same faculty, differing in degree only. For Hobbes, however, there is a distinction between imagining ('that is, having a certain idea') and conception ('that is, inferring by means of judgment that a certain thing exists'); and nominal definitions are an expression of the process of conceptual construction. See R. Descartes, *Philosophical Writings* tr. and ed. by E. Anscombe and P. T. Geach (London, 1975) for Hobbes's objections. According to the *OED*, the term 'conception' had been restricted to embryology until Hobbes appropriated it for philosophy.

[11] Colie, *Paradoxia epidemica* (note 1), provides numerous examples of these and other kinds of paradox.

reader to admire it and to wonder about it'.[12] We might conjecture, therefore, that paradox, for Harvey, was itself a means whereby other investigators—his ideal readers—would grasp, and then follow, the methodical path Harvey had taken and which he exemplified in his book. For although Harvey spent a lifetime investigating problems of generation, he could not provide definitive solutions to them as he had done for the circulation of the blood.[13]

In a recent interpretation of one of Harvey's strategies, Phillips Salman claims that Harvey used a quotation from Seneca's *Epistula* (58: 19–20) about the production of art, together with material from Aristotle's *Posterior Analytics* (100a5–9) on induction, to oppose the cognitive processes and methods of the artist to those Harvey wished to require for the physiologist.[14] I concur with Salman about the Seneca quote, for this presents the Platonic copy theory as follows:

> An Idea is an eternal exemplar of natural things. I will explain this definition that you may understand it the better. Suppose I intend to draw your picture. You yourself are the exemplar of that picture and from this exemplar my mind takes a representation which it imposes upon the work. So then it is your face which teaches and directs me and the imitation which is derived from it is the Idea.[15]

---

[12] Ibid., pp. 517–518.; see also p. 3, where Colie discusses the term, 'wondrer', coined by an Englishman in 1589 to describe the formal defense (a rhetorical paradox), 'organized along the lines of traditional *encomia*, of an unexpected, unworthy, or indefensible subject'.

[13] Ibid., for the paradoxes called '*insolubilia*'.

[14] P. Salman, 'William Harvey and art misplaced', *Annals of Science* (1992), 49: 3–19, who writes about the anatomist, not the physiologist.

[15] Salman's study, ibid., is a refreshing departure from traditional history of science interpretations of Harvey's work. But the author does not integrate statements in Harvey's preface with the rest of the text. Had he done so, he might have interpreted some portions of Harvey's strategy differently, as in the case of the Seneca quote which posits an 'eternal exemplar'. But which natural thing is to provide that exemplar in the case of embryology? Harvey's answer was that the state of affairs described by him in the case of the hen is the typical or paradigm case, so that mammalian and human embryology should be treated as special cases of the more general type exemplified by the hen. This answer is a paradox, of course, since it is contrary to the views of Harvey's predecessors, according to which viviparous not oviparous generation was the paradigm case.

This copy theory underpins the classical belief in an original or special creation of all individuals of a species, the term 'species' being understood in a Platonic sense as an eternally existing ideal form, pattern or exemplar of any class of things of which individuals in that class are imperfect copies and from which they derive their existence.[16] Since species pre-existed, each birth was conceived as an organisation *imposed from the outside* and explained by recourse to technical models from the visual arts.[17] To oppose such a belief, therefore, would require replacing those models with a different one; and, as we shall see, this is just what Harvey attempted.

In the case of Aristotle, however, I do not concur with Salman, for the passages from *Posterior Analytics* present the realist assumption that found its typical expression in the Platonic doctrine of ideas and that constitutes the metaphysical basis for Aristotle's theory of the universal in thought. Crudely put, when a number of individual eggs share a predicate of roundness, this is not because of a relation to something of the same kind as themselves but to something more ideal. This kind of realism may be illustrated with reference to two different aspects of Aristotle's thought.

On the one hand, the method which Aristotle called induction is that of seeing by intuition in a class of similar particulars the common features which constitute their essential nature, and then rising from this to the features common to various species which define their genus, and so on until we come to the necessary first principles of a science. Ascending by intuition is no less certain than scientific demonstration, so that we can descend again to particulars by syllogistic demonstration and our conclusions will remain necessary. Thus, Aristotle assumed that all things are what they seem and that we can generalise from what some things seem to what all things are.

[16] I.e., individuals are conceived as variations on a theme. For the twentieth-century 'new systematicists' opposition to this 'archetypal' or 'type' conception of species, see M. Beckner, *The Biological Way of Thought* (Berkeley, 1968), pp. 69–70.

[17] The Greeks distinguished between poets, whose art derived from divine inspiration, not knowledge, and craftsmen, whose artefacts derived from productive, not theoretical knowledge. See Tatarkiewicz, *A History of Six Ideas* (note 3), pp. 78–89.

On the other hand, although Aristotle followed the growth of ideas out of experience (and Harvey acknowledged this in his preface), these ideas—especially at the point where they are farthest removed from experience and immediate perception—are transformed by Aristotle from a logical product of human thought into an immediate presentment of a super-sensible world and the object, in that sense, of an intellectual intuition. And even when Aristotle intended a form to be something different from a universal, he treated form as more real than matter and, hence, as reminiscent of the sole reality of the Platonic Ideas.

Against Salman, therefore, I hold that in using the quote from Seneca and the material from Aristotle, Harvey's strategy is to set up direct realism for criticism, since that kind of realism was based on a science of being shaped after the pattern of geometrical forms that were static. And because Harvey's criticism develops as an attempt to emancipate thought from ontologism, this necessitates a redefinition of the universal in thought. Crudely put, if you see an egg and recall it by means of an image, you do not have a concept. But if you see a number of eggs and you think about roundness as a common attribute of all of them, you do have a concept. The object of your thought, in such a case, is a universal. But it is not a universal in the sense of the ancient realists, since, for Harvey, concepts are not found in pure intuition; rather, they are freely chosen conventions guided by experimental facts, but they must be defined synthetically in order to determine them precisely in logical thought. It is not surprising, therefore, that Harvey introduced the first new concept in embryology since Aristotle. This concept is the ovum.[18]

From the point of view of connotation, two attributes constitute Harvey's meaning of the ovum: first, the origin from moisture enclosed in some membrane or shell; and second, the mode of production whereby life manifests itself as a pattern of developmental stages up to that point when a univocal form emerges, universal for all species, whether plant, insect, bird or animal. From the point of view of denotation, the class of particulars belonging to the ovum are

---

[18] Harvey, *Disputations* (note 5), p. 424 *et passim,* who also utilised the term 'primordium' as a synonym for ovum. For the relation of Harvey's ovum concept to the modern one, see E. B. Gasking, *Investigations into Generation 1651–1828* (Baltimore, c.1966), pp. 27–28.

seeds, eggs of insects and birds, as well as the uterus of mammals.[19] Moreover, Harvey also suggested that the class of particulars includes the human brain, since it possesses both attributes of the ovum: just as the mammalian uterus is enclosed by two membranes (amnion enclosed in chorion), so too is the brain (pia enclosed in dura mater); and in both uterus and brain the mode of production generates a concept.[20]

You will note that Harvey's definition does not account for fertilisation. But Harvey could not solve this problem for two main reasons. First, the lenses of his magnifying glass were inadequate to see either spermatozoa or the female germ cell or the contact between the two.[21] Second, the problem seemed to involve an infinite regress of mechanistic explanation or explanation by efficient cause. All such explanations, however, have an arbitrary beginning, since causation must start from something, and wherever it starts no cause can be assigned for the initial datum. Failing to see the phenomena involved in fertilisation, Harvey's initial datum is the ovum, for this is where concept formation takes place.[22] Thus he could state: 'I call that

---

[19] The sex gland of the female was called 'testis'; it was given the name 'ovary' after Harvey's death. See C. McC. Brooks and H. A. Levey, 'Humorally-transported integrators of body function and the development of endocrinology', *The Historical Development of Physiological Thought*...ed. by C. McC. Brooks and P. F. Cranefield (New York, 1959), pp. 183–238.

[20] Harvey, *Disputations* (note 5), pp. 239, 349–350, 351, 445–446, several times asserted that conception in the brain is like conception in the womb. There is no evidence, however, that Harvey had read Plato's *Theaetetus*, in which Socrates (148e–151e *et passim*) practices the art of midwifery—i.e., argument by induction—to assist Theaetetus in bringing forth offspring—i.e., three definitions of knowledge, all of which are still born—i.e., in Greek, literally 'wind egg' or egg that fails to hatch.

[21] Spermatozoa were first seen in 1678; the mammalian egg in 1827; and the penetration of an egg by a spermatozoon in 1879. See Brooks and Levey, 'Humorally-transported integrators' (note 19), pp. 192–193 and Gasking, *Investigations into Generation* (note 18), p. 178.

[22] Although Harvey had much to say about causality, he relied on genetic analysis, not causal explanation when treating what happens in the ovum. Beckner, *The Biological Way of Thought* (note 15), p. 82, defines genetic analysis as follows:

> If we let E be the analysandum, a genetic analysis consists in exhibiting E as the outcome of a set of chains of events. Each chain, for the

fruitful or fertile which (unless some impediment happen from without) will reach its destined end by some efficient power implanted in it and accomplish that for the sake of which it was ordained'.[23]

Readers might do well, therefore, to interpret Harvey's book as an allegory about making, since, for him, 'every generation is a pathway to attaining form'.[24] But if his book is an allegory, then that allegory is a manifold, in which several levels operate at once. In what follows I shall select three levels only as central to Harvey's theory of creativity: first, how form is made; second, how designing is made; and third, how science is made. By means of this selection, I hope to draw attention to Harvey's attempt to provide a new technical model for the processes involved in making a new life as well as making a new science.

There is, however, one important point to note at the outset. In the case of making a new life, physiology is the theory that mediates Harvey's inferences and that informs his morphological concepts. This may seem an obvious point, when it is recalled that, because of his discovery of the circulation of the blood, Harvey is regarded as the founder of modern physiology. But it is still too little recognised that in all his works treating the blood, he insisted that functional concepts differ when the organism described is an embryo and when it is an adult. In the latter case, which will not concern us here, heart and blood work together as the unit of life; but in the former case, the blood alone is the unit of life for Harvey.[25]

> purposes of the analysis, is regarded as initiating with a single event, here called a beginner-event; each link in each chain is causally connected with the previous link, and is either contiguous with it or temporally overlaps it. A single event is the last link of all the chains (i.e., the links converge on one event); E is this last event, a spatiotemporal part of it, or a property of it or its spatiotemporal parts.

[23] Harvey, *Disputations* (note 5), p. 232.

[24] Ibid., p. 328.

[25] E.g., in the notes Harvey prepared for his Lumleian lectures of 1616, he wrote:

> blood [is] rather the author of the viscera than they of it. Because blood present before the viscera, not coming from the mother, but from a spot in the egg.

CREATIVITY AS PROCESS

1. HOW FORM IS MADE

In his preface Harvey acknowledged the industry of the ancients, while observing that they had not 'colonized' the truth, for there had been many notable discoveries 'lately' found out 'in the business of anatomy, not to mention the other arts'. But he believed that such discoveries had been made 'either by those who being wholly intent upon some one thing did by chance fall upon some other, or, which is the more commendable, by those who following Nature's lead with their own eyes along a winding but most sure path have at last attained to the very topmost peak of truth'.[26] Later in his book Harvey specifically developed the import of these statements, when he examined how one thing is made out of another, both in art and nature, 'and especially in the generation of animals'.[27]

According to Harvey, there are two different ways by which one thing can be made out of another. In one—let us call it fashioning—the artist proceeds like a sculptor, who prepares a liquid cement and, by applying this to 'the countenance of the dead' or to 'anything whatsoever', takes off an imprint or likeness. Then, when the imprint has grown hard, the sculptor fashions 'a mould' from which he 'cuts and divides the material that is already prepared and by paring away what is superfluous leaves the image remaining'.[28] In a second way—let us call it designing—the artist proceeds like a potter, who 'shapes the same image out of clay by adding to and augmenting ... so ... that at one and the same time he provides, prepares, fits and supplies the material (and in this way an image can be said to be made [i.e., generated] rather than fashioned)'.[29]

In the example of the sculptor, the material pre-exists, and form is embodied, first, as an imprint, next, as a mould and, finally, as an

---

See W. Harvey, *Lectures on the Whole of Anatomy*...tr. by C. D. O'Malley, F. N. L. Poynter and K. F. Russell (Berkeley, 1961), p. 90.

[26] Harvey, *Disputations* (note 5), p. 9.

[27] Ibid., p. 202.

[28] Ibid., pp. 68, 202.

[29] Ibid., p. 202.

image from the mould. But in the example of the potter, the material is both made and given shape at the same time. In natural science, according to Harvey, these two ways of making are named, respectively, metamorphosis and epigenesis, the differences between which may be stated briefly as follows: In metamorphosis, the starting point is from pre-existing material, and form is generated equivocally as larva to pupa to *imago*, whereas in epigenesis the starting point is 'a formative faculty', whereby form is generated univocally as a *conceptus*, 'like to itself in species'. In metamorphosis, each form is made entire at one time, whereas in epigenesis, form is made by successive accretions of parts that are given simultaneous increase and shape. In metamorphosis, form appears fully-grown and mature at birth, whereas in epigenesis, form increases in size and attains perfection after birth.[30]

In several passages Harvey remarked that chance also seemed to promote metamorphosis, as well as spontaneous generation from pre-existing putrescent matter.[31] But he afterwards noted that despite differences in the way forms are made, all making has two things in common. First, forms take their origin from an ovum; and this, Harvey wrote, is true regardless of whether that ovum 'derive from another of the same species or happen by chance to come from elsewhere'. Second, as in an egg, forms are made by a formative faculty, so too 'even in a fortuitous seed, there is a motive principle of generation, which generates out of itself and by itself'. For 'as it sometimes falls out in Art, so also in Nature', 'the same things sometimes happen by chance or accident which at others are accomplished by art', that is to say, by methodical design.[32]

Whilst acknowledging, then, that forms can be made by chance, by fashioning or by designing, Harvey asserted that only one way of

---

[30] Ibid., pp. 202–208.

[31] E.g., ibid., p. 293. Gasking, *Investigations into Generation* (note 18), pp. 30–31, points out that Harvey and his contemporaries

> always regarded the grub as one individual and the butterfly as another. The latter was thus considered a new organism generated within the pupa or Aurelia. Since the case did not increase in size, nor take in any new matter, the butterfly must be formed from the material present, therefore it must be an example of generation by metamorphosis.

[32] Harvey, *Disputations* (note 5), pp. 293–294.

making 'is properly called generation', namely, epigenesis, a term he first coined.[33] He therefore provided a number of different accounts of univocal concept formation. In one account he identified the earliest processes in both oviparous and viviparous generation, where there is, first, the ovum, produced by the female after intercourse and possessing the virtues of both parents; second, there is within the membranes of the hen's egg and the mammalian uterus, a clear liquid. Third, in that liquid, there appears a tiny red line, from the centre of which leaps a pulsating point. And fourth, from this pulsating point 'rays' of vessels take their origin.[34]

In another account, the fourth process marks the beginning of the generation of the vessels, which is followed by three subsequent processes. First, there appears a kind of 'concretion' which Harvey called 'a worm or maggot (for it seems to enjoy the life and obscure movement of a maggot)'. This coalesces into a jelly-like substance, the larger part of which consists of a rounded ball divided into three bladders. These bladders are the brain, cerebellum and eyes. Then, after 'the outline of the rest of the body takes shape', the liver, lungs, kidneys, cone of the heart and its ventricles and auricles, the guts and stomach, alimentary canal and genitalia all appear about the same time and take their origin from the veins and grow with them. At first, these viscera are white, bloodless and gelatinous until they take on their function. Finally, the skin is formed, along with all things made from it and that grow out of it.[35]

So far we have epigenesis or generation proper. But Harvey also compared epigenesis to nutrition, for he pointed out that during the formation of the vessels, the foetus, like a plant, is nourished 'through its roots', the umbilical and mesenteric veins. But when the

---

[33] Ibid., p. 203; see *A Supplement to the OED*, vol. 1. According to Harvey's translator, pp. lii–liii, it is not easy to decide whether Harvey accepted or rejected spontaneous generation (although he seems decisively to reject it, p. 424). But this problem disappears when his comparison between epigenesis and other forms of generation is understood as part of his strategy, the purpose of which will shortly appear.

[34] Ibid, pp. 427–432.

[35] Ibid., pp. 283–284. In process five 'all the parts spring from' the jelly-like substance, 'just as we see in automata … that go of themselves where one wheel sets the following wheel in motion'.

jelly is transformed into flesh and the motive organs are differentiated, the foetus is fed, like an animal, by sucking 'through the mouth'. In this description Harvey identified two ways of obtaining nutrition, the second of which enables the foetus to 'live the life of an animal', so that by repeated experience of sucking, the foetus not only grows but also acquires the habit of sucking. Indeed, Harvey remarked that a new born babe sucks 'with far more cunning and skill than a grown man', so that: 'It would seem ... to have learnt by long practice before it was born, for we know from experience by how short a discontinuance it unlearns it after birth'.[36] The implication is that knowledge (here as know-how) grows from repeated experience, a point Harvey had already made in his preface.[37]

2. How designing is made

From his own experience, including repeated dissection, Harvey concluded that life develops spontaneously within the ovum and that the blood, not the heart, is the unit of life, for 'according to the law of Nature', the blood exists in the ovum before the pulse as well as before its receptacles. As the principal part, the blood is the origin and designer of all the rest, which it uses as its instruments. Hence, in the blood resides both 'the beginning, from which comes movement, and the end', from which movement ceases.[38] In these remarkable

---

[36] Ibid., pp. 306–307.

[37] Ibid., pp. 12–13, where Harvey wrote:

> as Art is a habit concerned with doing, so is Knowledge a habit concerned with comprehending ... The source of both is from sense and experience ... [but] ... without the right verdict of the senses controlled by frequent observations and valid experience, we make judgments entirely on phantoms and apparitions inhabiting our minds.

[38] Ibid., p. 283, implying the paradox: in my beginning is my end. See also, ibid., p. 207: 'from wheresoever comes the first drop of blood in the egg, from there also comes by way of generation the whole mass of the chick, and from this blood are constituted the members or organs of its body; and likewise all the homogeneous parts, as skin, flesh, vein, membrane, sinew, cartilage and bone, derive their origin from this blood'. But, p. 254, the blood is homogeneous only as living principle (it is heterogeneous as gore).

statements, Harvey rid embryology of the dualism between matter and form, thereby making his most notable break with Aristotelian tradition.

That tradition, as well as the tradition of the Galenic physicians, relied on Aristotle's assumption that the matter of menstrual origin in the uterus is fashioned into a form by the male semen in the same way that a sculptor acts on material to produce a piece of statuary.[39] In this technical model, form is embodied from pre-existing material 'concocted' by chance; it is abstracted by removing superfluous material; and it is complete once the material has been subtracted. Hence, Aristotle's theory is what Harvey described as metamorphosis in contradistinction to epigenesis or generation proper. Here, then, is the reason for Harvey's comparison between making by fashioning and making by designing, since his description of epigenesis as a simultaneous process of growth and differentiation cannot be likened to a sculptor acting on material.[40]

A closer analogue might be a potter whose art depends for its efficacy upon the physical properties of the clay that is worked and which for the time being is essentially a fluid. For form to arise from this material, however, there must be a constant and uniform pressure exerted by the potter's thumb, since only then will the resulting form display surfaces of equilibrium as manifested under certain limiting conditions. These conditions, or lines of constraint, determine the action of the expansive forces of differentiation and growth that would otherwise be uniform and symmetrical. In short, the potter's art exhibits forms that a fluid surface can assume under surface tension (capillarity).[41]

---

[39] See, e.g., G. E. R. Lloyd, *Polarity and Analogy: Two Types of Argumentation in Early Greek Thought* (Bristol, 1987), pp. 285–286 *et passim*.

[40] For the Aristotelian tradition, as well as that of the Galenic physicians, see Gasking, *Investigations into Generation* (note 18), pp. 16–36, especially pp. 33–34, where she points out, with evidential statements, that Harvey disagreed with Aristotle on all the important issues, thereby correcting such interpreters as W. Pagel, *William Harvey's Biological Ideas*...(Basel, 1967).

[41] For surface tension, see D. W. Thompson, *On Growth and Form*...(abridged, Cambridge, 1994), pp. 43–44, 49–87, who points out, p. 69, that the potter's art illustrates some complex biological problems of 'a figure of equilibrium which is an *open* surface of revolution'.

Although Harvey himself referred to the potter's art as an example of epigenesis, he recognised that that art cannot provide an adequate representation of making by designing, because it retains one of the main features of Aristotle's technical model, namely, the distinction between artist and artefact. This distinction implies that the artist works material through the medium of his tool, whether that tool is the hand or an instrument held by the hand, whereas in Harvey's embryology, matter works itself, because the blood is simultaneously artist and artefact, designer and design. But lines of constraint determine the blood's dynamic activity, otherwise no form would emerge that is universal for all species. Harvey, therefore, resorted to a different model, when he described the blood as moving and dancing like an animal.[42]

The impulse to dance is revealed when, in the ovum, in a tiny line of blood, almost in its centre, 'there leaps a capering bloody point ... that ... flashes like the smallest spark of fire'.[43] From that moment, the dance of life begins, for in addition to the blood's harmony, or pulsation, there is 'a wavering and perpetual motion' of growth and differentiation and, after the receptacles are formed, a rhythm of circulation. Moreover, the dance danced may vary in speed and intensity, depending on different circumstances, for Harvey's experiments led him to believe that the blood is irritable and, hence, sensitive—just as fire 'like a living thing exists in continual flux'.[44] Indeed, since fire and blood are both 'like an animal [that] moves

---

[42] I.e., in rhythm and harmony, concepts that Harvey first applied in connection with his studies of muscles, for which see Chapter 5 *infra*; see also W. Harvey, *De motu locali animalium 1627* ed. and tr. by G. Whitteridge (Cambridge, 1959), pp. 67, 111, 143–149, 153, where he stated, p. 145: 'Nature ... attains her end by means of rhythm and harmony'; and W. Harvey, *An Anatomical Disputation concerning the Movement of the Heart and Blood in Living Creatures* [*De motu cordis 1628*] tr. by G. Whitteridge (Oxford, 1976), pp. 39, 50–51.

[43] Harvey, *Disputations* (note 5), p. 96.

[44] Ibid., p. 389; see also p. 100 (concluded from an experiment), where Harvey noted that the bloody point, 'by an oft-repeated touch ... becomes confused in the rhythm and order [i.e., harmony] of its pulsation', and p. 247, where he pointed out that 'from the variations in its [the blood's] motion, in its speed or slowness, its vehemency or weakness etc., it is plain how sensitive it is'.

itself, nourishes and increases', flame may be said to be 'the symbol of life'.[45]

As long as the blood lives, that is to say, performs its dance, the creature made from it lives. What, then, represents the constraints of the dance? Harvey, of course, could not answer this question, although he hinted that the 'law' or proportional relation of 'the first efficient' is where the concept ('ratio') resides.[46] But his technical model implies that a particular measure is assigned to the dance of each type of creature and that this measure constrains the spatiotemporal process.[47] Of course, the spatiotemporal process is incompletely understood even today, for geneticists, working within the metaphor of a genetic code, have yet to explain how a one-dimensional linear model can account not only for form as three-dimensional space but also for processes in time.[48]

Given this difficulty, Harvey's attempt to introduce a new technical model deserves our appreciation, for by conceiving the blood as a harmony and rhythm in space and in time, he could stress continuity between the vital activities of the ovum, the foetus and the mature animal. Unfortunately, however, there has been little appreciation of Harvey's achievement, even though today the earliest stages in embryonic development are often described in terms of the

---

[45] Ibid., p. 387.

[46] Ibid., p. 232. For the general principle underlying the theory of surface tension, see Thompson, *On Growth and Form* (note 40), p. 88.

[47] The English term, 'measure', as referring to geometrically patterned dances and dancing in general, was current from c.1550–c.1650. For some of these dances, see J. M. Ward, 'Newly devis'd measures for Jacobean masques', *Acta Musicologica* (1988), 60: 111–142; see also M. Dean-Smith, *Playford's English Dancing Master 1651*...(London, 1957), p. xviii, who states: 'an infinite variety was achieved by permutations of simple, constant elements, with the [geometrical] figure or figures peculiar to a particular dance distinguishing it from others'.

[48] The genetic code is protean: life in its many forms is basically the same. Two metaphors for the code have been common—one from architecture (a blueprint, i.e., pre-existing molecules as building blocks, copy as building); the other from cookery (a recipe rather than a blueprint). For other metaphors from control and communication theory, see E. F. Keller, *Refiguring Life: Metaphors of Twentieth-Century Biology* (New York, 1995).

dance.[49] In 1926, for example, one finds reference to 'The Dance of the Chromosomes'.[50] And in 1968 one finds an embryologist who preferred the term, 'morphochoreisis', instead of gastrulation to designate 'the mobilisation of cellular elements, their arrangement in groups of distinct destiny, all in obedience to the severe rules of some ceremonious dance, typical for each type of creature'.[51]

3. How Science Is Made

From the foregoing, we may conclude that, for Harvey, concept formation is a methodical process. In embryology, that process is the outward and visible sign of the blood's dynamic activity, the stresses and strains of which force apart and draw together, thereby systematically assembling from its own substance different shapes and textures, each of which specialises into something helpful to the whole. Not only is the whole specific but also a particular individual within the limits of the specific. Thus arises the new coherent individual to be, so that it seems *as if* an immanent rational principle

---

[49] E.g., Gasking, *Investigations into Generation* (note 18), describes Harvey's achievement in negative terms, when she points out that his discoveries 'disproved, and correctly disproved, all the previous theories, and rendered untenable any model for fertilisation which involved the totality of the semen and an equivalent menstrual clot'. This is to ignore the 'pregnant' suggestions in Harvey's book: e.g., the ovum, like the nucleus of a cell, is a seat of energy. For the implications, see Thompson, *On Growth and Form* (note 41), p. 13, who writes:

> in the attractions or repulsions of the parts of the dividing nucleus and in the 'caryokinetic' figures which appear in connection with it, we seem to see in operation forces and the effects of forces which have, to say the least of it, a close analogy with known physical phenomena.

The term, karyokinesis, refers to the series of active changes taking place in the nucleus of the living cell in the process of division.

[50] M. C. Stopes, *The Human Body* (London, 1926), p. 104, referring to karyokinesis.

[51] A. M. Dalcq, 'Form and modern embryology', *Aspects of Form*...ed. by L. L. Whyte (London, 1968), pp. 91–120, pp. 100–101, referring to gastrulation, when a blastula, or hollow sphere formed of a single layer of cells, is converted into a gastrula, or two cellular layers and a primitive gut.

inspired the blood with information for carrying out a design. Yet, Harvey insisted that in the works of nature,

> there is neither prudence, nor skill, nor understanding, but they seem such only to our notions who judge of the divine productions of Nature by our own arts and faculties (as being copies drawn after them by our own selves), *as if* the active principles of Nature did produce their effects in the same way in which we do our own works that we have made by art, that is by judgement and the discipline acquired by the exercise of the understanding mind.[52]

The making of a *conceptus*, therefore, cannot be accounted for by the Platonic doctrine that Harvey illustrated by a quotation from Seneca with reference to art, namely, that the *eidola* or image is what is seen in the artefact, whereas the *eidos* or species is in the mind of the artificer and, hence, quite outside the artefact and not only outside it but in being before the artefact was made. Nor can the making of conceptions be accounted for by the Aristotelian doctrine in the *Posterior Analytics* that Harvey summarised with reference to the acquisition of knowledge. For by assuming that there is only one way of seeing and copying an object as it is, both Platonic and Aristotelian doctrines fail to account not only for the active principles of nature but also for the active, interpretive role of the imaginer.

Although Harvey accepted that the sense image is a kind of universal that differs from the mind's awareness of a sensation, he rejected the classical belief that this kind of universal can provide the first principles of science. To illustrate why, Harvey put forward a hypothetical example in which poets and painters,

> though they behold one and the same object at the same moment and in the same place, all other circumstances being alike, yet will each one of them express and describe it in a different way in accordance

---

[52] Harvey, *Disputations* (note 5), p. 236, italics mine. Harvey frequently states that a *conceptus* seems to be the end result of consideration and hard thought, that its wonderfulness is a designed or artful wonder. But at the root of this statement is the 'wondrer' paradox, utilised by Harvey with great skill (see note 11 above). Although Salmon, 'William Harvey' (note 13), pp. 10–11, may be the first to have drawn attention to Harvey's use of wonder; his definition of wonder as a disinterested search for causes follows that of Thomas Aquinas.

> with the several ideas which they have each drawn of it in their imagination ... the reason for ... this [being] that in vision or in the act of seeing, each particular by itself was clear and distinct; but the object being removed (as if, for example, you shut your eyes), the same particular is abstracted in the imagination or laid up in the memory and appears obscure and confused, nor is it any longer apprehended as a particular but as some general and universal thing.[53]

According to Harvey, obscure and confused universals are preconceptions, because they are generated from principles that 'originate from things perceived by the senses'. To form true universals or conceptions, we must take 'special care that those principles are safely grounded' in a methodical process that involves repeated vivisectional, anatomical, pathological and embryological observation, as well as repeated experimental demonstration.[54]

These statements indicate Harvey's indebtedness to the Stoics, who distinguished between preconceptions and conceptions by treating the former as an undeveloped conception that comes about by chance and the latter as a thought-out definition that comes about methodically by design.[55] They also distinguished between nature, the divine life principle, and convention, the nature we make ourselves. Convention is our art, which the Stoics defined as a habit of creating in a methodical way, thereby modifying Aristotle's definition of art as a disposition for making with correct reasoning. For the Stoics, as for Harvey, the true human artist does not depend on chance or trial and error but makes use of a regular process to bring something into existence. And for the Stoics, as for Harvey, the norm for this process is creative nature.[56]

---

[53] Harvey, *Disputations* (note 5), p. 11.

[54] Ibid., p. 13.

[55] See F. H. Sandbach, 'Ennoia and prolepsis in the Stoic theory of knowledge', *Problems in Stoicism* ed. by A. A. Long (London, 1971), pp. 22–37; see also E. V. Arnold, *Roman Stoicism* (London, 1958), pp. 135–141. Preconceptions make possible the search for, and discovery of, new knowledge that is more definite.

[56] See D. E. Hahm, *The Origins of Stoic Cosmology* (Columbus, Ohio, 1977). For the Stoics, nature is a craftsmanlike fire proceeding methodically—literally, by a path—to genesis. Such was Zeno's definition; and such, too, was Harvey's,

In his preface, therefore, Harvey advised readers to 'take on trust nothing that I say of the generation of animals'; rather, embrace 'the practice of viewing Nature herself with your own eyes'.[57] The rest of Harvey's book develops the implications of these statements, namely, that scientific constructions are erected not so much upon first principles that are immediately suggested by simple observations, as upon principles which always follow as consequences from the preceding scientific development. For what we deduce from provisionally accepted hypotheses leads through experimental verification to their criticism and renovation. Thus we are to understand the conclusion to Harvey's preface, where he wrote:

> Be not offended ... if ... I make use of a new method and sometimes of unusual words, and do not think me thereby more desirous of vain-glory than of advantaging others by true observations grounded in Nature's self. To prevent this, know that in this matter I tread but in the steps of other men who have lighted me the way, and, so far as is fit, I make use of their words.[58]

In place of the classical science of being, therefore, Harvey proposed a new science of becoming.

CONCLUSION

Harvey's book was published with a frontispiece showing Jove holding two halves of an egg out of which spills life in its different forms. The engraving thus depicts emblematically the motto of the book—*Ex ovo omnia* (all things from an egg), where the ovum is the formless space within which the dance of life first manifests itself. But Harvey could not explain how the ovum obtained 'the power of effectuating something'; that is, he was unable to discover the

---

Disputations (note 5), pp. 381, 387, 389, since in the epigenetic process, the blood or life principle, dances by a methodical path like a living flame.

[57] Ibid., p. 13.

[58] Ibid., p. 20.

efficient cause of the beginning of motion.[59] And even after his book was published, Harvey continued to worry about this problem, as is clear from an annotation that he made to a printed copy of his book, where he wrote: 'see Bacon, On difficulties, at the beginning'.[60]

Although Harvey's note is cryptic, it may refer to the essay of Francis Bacon, 'On principles and origins'[61], in which we find this allegory for natural philosophers: Cupid, beginning from an egg laid by Nox, has no parent (efficient cause), God always excepted. A natural philosopher should continually recollect that Cupid has no parent, lest his understanding turn aside to universal conceptions that are unrealities.[62] Next to God, therefore, Cupid is the cause of causes.

---

[59] Ibid., p. 448, referring to fecundity—another starting-point problem. Harvey reminded his readers that a number of conjectures had been tried to solve this problem. Some natural philosophers, for example, refurbished the ancient opinion concerning atoms, whereby a 'most subtle effluvia coming from the semen of the male and emanating freely, after the manner of odours', is shot into the uterus at the time of coitus. Others resorted to incorporeal spirits 'like so many demiurges', whereas still others had understood fecundity on the analogy of 'a ferment or a leavening'. Harvey, therefore, 'invented' a conjecture in order 'to stir the wits of the studious to enquire more deeply into the matter'. According to his conjecture, fecundity might be transmitted by 'contagion', that is, by action at a distance without any apparent material contact, just as diseases are spread or the way in which iron is attracted to the magnet or the way in which strings tuned in the same tension vibrate in sympathy one with another. See J. C. Kassler, *Inner Music: Hobbes, Hooke and North on Internal Character* (London, 1995), pp. 68–69, 240–241 n.92.

[60] Harvey, *Disputations* (note 5), p. 457.

[61] For the essay, left incomplete, see F. Bacon, 'De principiis atque originibus [c.1610–20]', *Philosophical Studies*...tr. by G. Rees and M. Edwards (Oxford, 1966). See also F. Bacon, 'Cupid, or an Atom', *Wisdom of the Ancients* [1625] (London, 1907), pp. 238–242, which seems to recast some of the material in the former essay, thereby emphasising the need to investigate beginnings of motion.

[62] In Bacon, 'Cupid, or an Atom' (note 61), the 'unrealities' are those of the ancient realists: Plato, who made over the world to thoughts, and Aristotle, who made it over to words. For the source of Bacon's allegory in Orphic cosmogonies, see G. S. Kirk, J. E. Raven and M. Schofield, *The Presocratic Philosophers: A Critical History with a Selection of Texts* (2d edn., Cambridge, 1983), pp. 17–23, 33. See also E. Darwin, *Zoonomia*...(2 vols., London, 1796), vol. 1, p. 533, who wrote:

> This idea of the gradual generation of all things seems to have been familiar to the ancient philosophers as to the modern ones; and to have

As such, the cause of causes is positive, but inexplicable—it must be taken absolutely as found and must not be judged by any previous conception. The same message appears in Harvey's allegory about making, which begins from an ovum. Regardless of whether the ovum is seed, egg, uterus or brain, its unit of life, though uncaused, is the parent of the effect and, therefore, the cause of causes in the ovum, God always excepted.

---

given rise to the beautiful hieroglyphic figure of the … first great egg, produced by NIGHT, that is, whose origin is involved in obscurity, and animated by … DIVINE LOVE; from whence proceed all things which exist.

5. ON THE STRETCH
HOBBES, MECHANICS AND THE SHAKING PALSY

... it is not when parts are more relaxed than usual that spasms and tetanus supervene, but when they are more on the stretch
(Hippocratic writings 'On Fractures')

INTRODUCTION

In 1647 the much-maligned English philosopher, Thomas Hobbes (1588–1679), had a protracted illness, which he afterwards described briefly in a letter and from which description we may identify four stages. First, there was a painful and continuing fever, during which period Hobbes was delirious. This lasted six weeks. Next, as the fever waned, abscesses (*apostemata*) broke out, so that he was confined to bed for four more weeks. Then, after the abscesses healed, he suffered excruciating pain, which he attributed to sciatica (*ischiadica*). Indeed, the attending physician recorded that the pain was such that Hobbes wished to kill himself. And finally, there was apparent recovery, for as the pain became milder, the will to live reasserted itself.[1]

Shortly after this illness Hobbes's hands began to tremble, thus manifesting the first symptom of a 'shaking Palsey'. According to his friend and biographer, John Aubrey, the tremor 'began in France

[1] For the letter, written in Latin to Samuel Sorbière and dated 27 November 1647, see G. Croom Robertson, 'Some newly discovered letters of Hobbes', *Mind* (1890), 15: 446–447. N. Malcolm (ed.), *The Correspondence of Thomas Hobbes* (2 vols., Oxford, 1994), vol. 1, pp. 163–164, translates *ischiadica* as swelling of hip joints. See also G. Patin, *Letters*...ed. by J. H. Reveillé-Perise (new edn., 2 vols., Paris, 1846), vol. 2, pp. 593–594. Apparently, Patin treated Hobbes for several illnesses, according to F. R. Packard, *Guy Patin and the Medical Profession in Paris in the XVIIth Century* (New York, 1970), p. 89.

before the year 1650, and 'haz growne upon him by degrees, ever since, so that he haz not been able to write very legibly since 1665 or 1666, as I find by some letters he hath honoured me withall'.[2] By 1663, according to another friend, Robert Hooke, Hobbes's hands 'shook as fast one way as his head did the other'.[3] And Aubrey also reported that Hobbes was 'for severall yeares before he died so Paralyticall that he wase scarce able to write his name, and that in the abscence of his Amanuensis not being able to write anything, he made Scrawls on a piece of paper to remind him of the conceptions of his Mind he design'd to have committed to writing.'[4]

From this evidence, we may conclude that Hobbes's shaking palsy followed the pattern first described in detail by James Parkinson, after whom the syndrome now is named.[5] If we adapt Parkinson's description to the case in question, we may gain some insight into Hobbes's experience, which would begin with a slight sense of weakness and a proneness to tremor in one of his hands and arms. Although these symptoms would have gradually increased in the part first affected, they would have been felt in some other part, presumably the head, within less than a year. Not long afterwards Hobbes would have found difficulty in preserving an upright posture; and as this symptom increased, one and then the other of his legs would have been affected with tremor and loss of power.

Up to this point, Hobbes would have experienced little inconvenience because of the strong influence of habitual endurance. Soon, however, the performance of skilled tasks would require increased attention and greater effort. Being harassed by a tormenting round of tremor, Hobbes would seek relief in exertions such as

---

[2] J. Aubrey, *Brief Lives* ed. by O. L. Dick (Harmondsworth, 1978), p. 315.

[3] (Letter to) R. Boyle, *The Works* ed. by T. Birch (6 vols., London, 1965-66), vol. 6, p. 486. For this letter and evidence of Hooke's friendship with Hobbes, see J. C. Kassler, *Inner Music: Hobbes, Hooke and North on Internal Character* (London, 1995), p. 128.

[4] Aubrey, *Brief Lives* (note 2), p. 316.

[5] J. Parkinson, *An Essay on the Shaking Palsy* (London, 1817), pp. 3–9, who did not mention such symptoms as immobile facial expression, eye movement abnormalities and loss of strength in the modulation of the voice. See L. J. Findley and R. Capildeo (eds.), *Movement Disorders: Tremor* (London, 1984), who identify three cardinal features of Parkinsonism: tremor, rigidity and bradykinesia.

walking or tennis, for these would divert his attention from the unpleasant feelings. But, eventually, even this temporary relief would be denied, because Hobbes would develop a propensity to lean forward, thus being forced to walk on tip-toe and, at the same time, irresistibly impelled to take much quicker and shorter steps, and thereby to adopt unwillingly a running pace. At last, the trunk of Hobbes's body would become almost permanently bowed, the influence of the will over the muscles would decay little by little, and tremor would become so vehement as to make sleep impossible. Since the disease rarely kills, release would come from other causes, which in Hobbes's case was a stroke.[6]

To understand the disorders of Parkinsonism, one needs to understand normal movement to see where the strategy has fallen down. Such an understanding, I believe, is to be found in Hobbes's neglected philosophical masterpiece, which was partially written before 1647 and was subsequently completed and published, first in Latin in 1655, and then in English in 1656, under the title, *De corpore, Concerning Body*.[7] In this work Hobbes sought to describe nature by recourse to the concept of continuity. In so doing, he rejected the species theory of the Peripatetics, the continuous atomism of Descartes, and the emission theory of the neo-Epicureans, for all these theories described nature in terms of recognisable individual elements, like the numbers 1, 2, 3, ...[8] But individualised elements, like numbers, are not the appropriate image for the concept of continuity.[9] For this concept we require an image like a string stretched in three dimensions with resistance.

---

[6] See S. Mintz, 'Thomas Hobbes', *Dictionary of Scientific Biography* ed. by C. C. Gillispie (16 vols., New York, 1970–80), vol. 6, pp. 444–451.

[7] T. Hobbes, 'Elementorum philosophiae...de corpore [1655]', *Thomae Hobbes Malmesburiensis Opera philosophica quae Latine scripsit* ed. by W. Molesworth (5 vols., London, 1839–45), vol. 1; T. Hobbes, 'Elements of philosophy...concerning body [1656]', *The English Works of Thomas Hobbes of Malmesbury* ed. by W. Molesworth (11 vols., Scientia Aalen, 1962), vol. 1.

[8] See Kassler, *Inner Music* (note 3), pp. 49–51, 76–79.

[9] For continuity as 'no nextness', see E. T. Bell, *Mathematics: Queen and Servant of Science* (New York, 1951), pp. 297–298, 301–302, and *Men of Mathematics* (New York, 1961), p. 13. See also M. J. White, *The Continuous and the Discrete: Ancient Physical Theories from a Contemporary Perspective* (Oxford, 1992), who does not include the arguments of music theorists, i.e., the

The Stoics had used this image of a stretched string;[10] but Hobbes reinterpreted the Stoic image, for he conceived functions immanent in organisation like bundles of strings of different lengths and tensions. He then developed this conceptual model by drawing on the new science of music that emerged in the early part of the seventeenth century from problems relating to musical strings, elastic substances that vibrate.[11] It is noteworthy, therefore, that during his third residence in France, which extended to eleven years, Hobbes paid daily visits to Marin Mersenne, who by 1636 had established the chief laws of the stretched string: that the frequency of the string is directly proportional (a) to its length, (b) to the square root of the weight which stretches it, (c) to the square root of the weight of the string itself, (d) to the reciprocal of its diameter, and (e) to the reciprocal of its specific gravity.[12] And in 1644 Mersenne published a collection of tracts, in which he expounded some of Hobbes's ideas.[13]

Pythagoreans who favoured the discrete and the Aristoxenians who favoured the continuous.

[10] See D. E. Hahm, *The Origins of Stoic Cosmology* (Columbus, Ohio, 1977), pp. 3–10.

[11] For details, see Kassler, *Inner Music* (note 3), pp. 49–107.

[12] M. Mersenne, *Harmonie universelle*...(Paris, 1636), whose experiments are described by F. V. Hunt, *Origins in Acoustics: The Science of Sound from Antiquity to the Age of Newton* (New Haven, 1978), pp. 89–94. See also S. Dostrovsky, 'Early vibration theory: Physics and music in the seventeenth century', *Archive for History of Exact Sciences* (1974–75), 14: 169–218.

[13] M. Mersenne, 'Ballistica' and 'Hydraulica', *Cogitata physico mathematica*...(Paris, 1644), contains expositions of Hobbes's ideas as follows: 'Ballistica, Praefatio...ad Lectorem' (unpaginated, on the faculties of the soul); 'Ballistica, Corollarium post Prop. XI' (pp. 29–30, on dynamics and 'reditus arcus'); 'Ballistica, Proposition XXIV et Monita I et II' (pp. 74–82, on theory of light and planetary motion); 'Hydraulica, Corollarium II post Propos. XXV' (pp. 129–131, on 'De parabola helici Archimedeae aequali'). See also J. Wilkins, 'Mathematical magic [1648]', *The Works*...(2 vols., London, 1802), vol. 2, p. 162, who wrote

> He that would be informed in the nature of bows, let him consult *Mersennus de Ballistica et Acontismologia*, where there are diverse subtile enquiries and demonstrations, concerning the strength required to the bending of them to any distance, the force they have in the discharge, according to several hints, the strength required to be in the string of

De corpore, then, is a theory of mechanics; but it is a practical theory, for it presents a methodical procedure for investigating motion. This procedure is threefold, for first, 'we are to search out the ways of motion simply'; next, 'the ways of such generated motions as are manifest'; and lastly, 'the ways of internal and invisible motions'.[14] The initial study is that of geometry, which treats ideal motion or motion in the abstract. The second study is kinematics, which deals with the motion of particles and bodies. And the third study is that of dynamics, which is concerned with forces and how they affect the kinematics. According to Hobbes, only the last study is 'the enquiry of natural philosophers'.[15] And in the epistolary dedication to *De corpore*, he observed that, in England, 'the only true natural philosophers' are physicians, especially 'our most learned men of the College of Physicians in London'.[16]

You will note that Hobbes's method proceeds inward, like the great anatomy of Andreas Vesalius, who, in *De humani corporis fabrica libri septem* (1543), stripped the human creature layer by layer in an investigation that also was a method of discovery.[17] Moreover, this stripping away was depicted in a number of magnificent illustrations, some of which revealed bundles of strong cords stretched tight, like the tuned strings of the lutes illustrated in *Figure 5.1*.[18] In the early 1640s, in Paris, Hobbes read Vesalius's book

---

them, the several proportions of swiftness and distance in an arrow shot vertically, or horizontally, or transversally.

For Hobbes's contribution to Mersenne's tracts, see below (note 21).

[14] Hobbes, 'Concerning body' (note 7), p. 73.

[15] Ibid.

[16] Ibid., p. ix.

[17] According to R. L. Colie, *Paradoxia epidemica: The Renaissance Tradition of Paradox* (Princeton, 1966), p. 431, the subject of Vesalius's book provided a metaphor for all sorts of 'uncoverings', in which the investigators sought to articulate, as anatomy does, the disparate parts of a system or systems into a fitting whole.

[18] According to Mintz, 'Thomas Hobbes' (note 6), Hobbes owned a bass viol. But Kassler, *Inner Music* (note 3), p. 59, suggests that Hobbes also played the lute, an inference drawn from the musical inventory of Hobbes's library provided by L. Hulse, 'Hardwick MS 29: A new source for Jacobean lutenists', *The Lute* (1986), 26: 63–72.

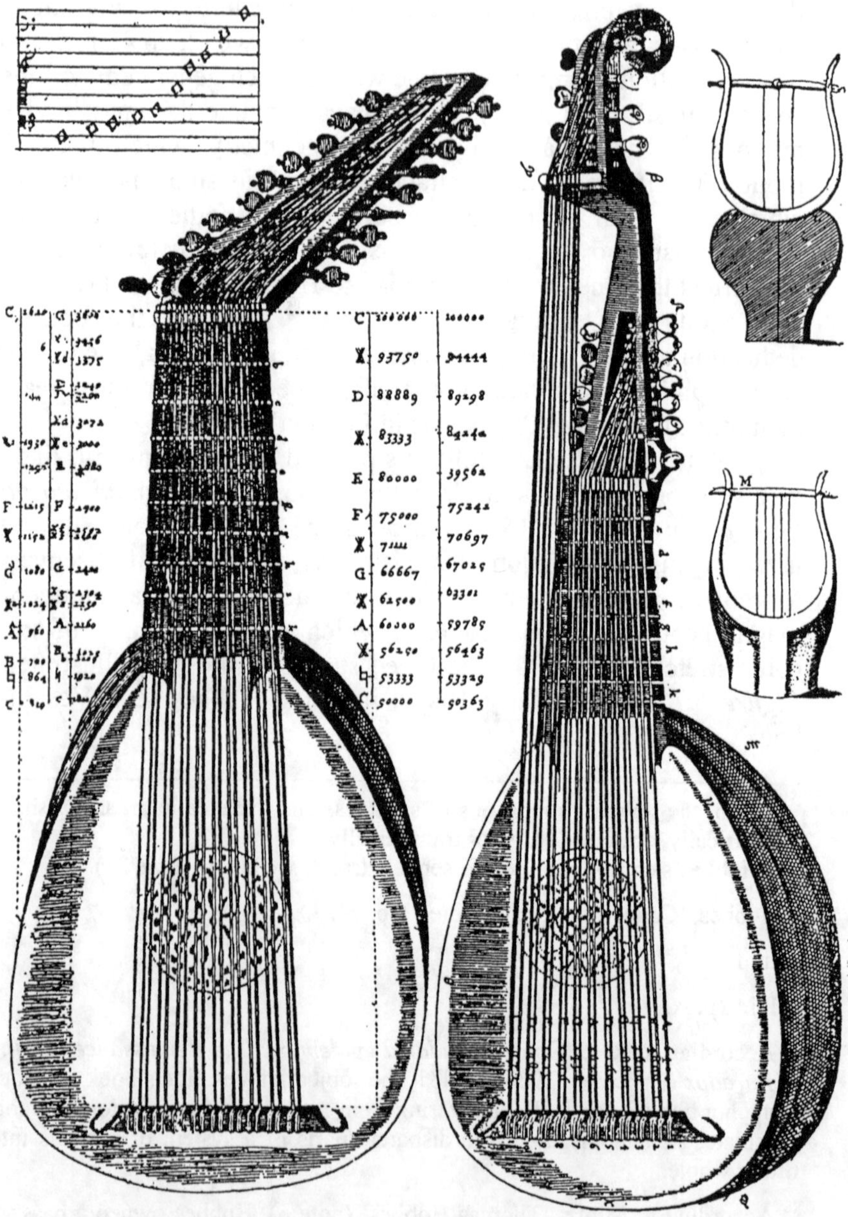

Figure 5.1. Lute (left) and double-necked lute (right), M. Mersenne, *Harmonie universelle...*(Paris, 1636)

with his friend, William Petty,[19] and it is possible that during this period the idea took root that the cohesion of body—its holding strength and degrees of resistance—could be explained by recourse to stretched strings of different lengths and tensions.

That Hobbes was investigating problems of cohesion appears from two tracts written before and after his reading of Vesalius. In the first tract, *De mundo dialogi tres* (1642), he attributed the cohesiveness of body to the resistance of its parts to any exterior motion tending to destroy it and explained the motions of the earth as the effect partly of wave fronts created by the contraction and dilation of the sun and partly of the earth's turning like an animal before a fire, so that each section successively receives solar heat.[20] And in a second tract, now known as *Tractatus opticus* (1644), he argued that body has degrees of resistance that affect the kinematic propagation of physical rays from the contracting and dilating sun; and he conceived physical rays as infinitesimal portions of an expanding wave front.[21]

The term, resistance, utilised in both tracts, implies cohesive force and, hence, is an 'enquiry of natural philosophers'. It is noteworthy, therefore, that Hobbes's interest in such enquiries predates the 1640s, as appears from his friendship with the physician, William Harvey.[22] Although the two men became close friends during

---

[19] Aubrey, *Brief Lives* (note 2), p. 399: 'At Paris he [Petty] studied Anatomie, and read Vesalius with Mr. Thomas Hobbes, who loved his company. Mr. H then wrote his *Optiques* [see below, note 21]; Sir W. P. then had a fine hand in drawing and limning, and drew Mr Hobbes Opticall schemes for him, which he was pleased to like.'

[20] T. Hobbes, *Critique de 'De mundo' de Thomas White* [1642] ed. by J. Jacquot and H. W. Jones (Paris, 1973), which includes, besides the edition of Hobbes's manuscript, his Latin poem on the motion of the earth and his English notes on an early draft of some chapters of *De corpore*; see also T. Hobbes, *Thomas White De mundo examined*...tr. by H. W. Jones (London, 1976).

[21] T. Hobbes, 'Opticorum libri septem Praefatio ad Lectorem IV', M. Mersenne, *Universae geometriae, mixtaque, synopsis et bini refractionum demonstratum tractatus* (Paris, 1644), pp. 472–475. For a detailed treatment of Hobbes's treatise and its reception, see A. E. Shapiro, 'Kinematic optics: A study of the wave theory of light in the seventeenth century', *Archive for History of Exact Sciences* (1973), 11: 143–172, who is the first to point out that Hobbes began the kinematic tradition in continuum theories of light.

[22] Aubrey, *Brief Lives* (note 2), pp. 292, 318.

the last years of Harvey's life, when Hobbes was completing *De corpore*, their first meeting seems to have occurred between c.1621 and c.1626, when Hobbes was amanuensis to Francis Bacon, one of Harvey's patients. Harvey's researches on muscle and on animal movement were carried out between c.1616 and 1628, and there are striking similarities between some portions of Harvey's draft treatise on animal movement and Hobbes's chapter on the senses in *De corpore*.[23] From internal evidence, that chapter appears to have been written earlier than other portions of the treatise, some of which are developments from Harvey's book, *Exercitationes de generatione animalium* (1651).[24]

Harvey commenced writing that book in the early 1630s, when he also performed dissections on the King's deer.[25] During this period Hobbes witnessed Harvey's dissections, which were vivisections—in the tradition of Vesalian anatomy. Indeed, Hobbes recorded that:

> At the breaking up of a Deer I have seen it [pulsating motion] plainly in his Bowels as long as they were warm. And it is called the *Peristaltique* Motion, and in the Heart of a Beast newly taken out of his Body; and this Motion is called *Systole* and *Diastole*. But they are both of them this compounded [i.e., pulsating] Motion, whereof the former causeth the food to Winde up and down through the guts, and the latter makes the Circulation of the Blood.[26]

The gastrointestinal tract and the heart are both cordlike substances that pulsate. Although the sensory 'image' of cords pulsating may suggest a lawlike regularity in nature, the image must be brought to certainty by testing through experience. For this, Hobbes could not

---

[23] G. Keynes, *The Life of William Harvey* (Oxford, 1966). For Harvey's treatise, see below (note 51).

[24] See Kassler, *Inner Music* (note 3), pp. 66–69.

[25] R. G. Frank Jr., *Harvey and the Oxford Physiologists: Scientific Ideas and Social Interaction* (Berkeley, 1980), pp. 38–42.

[26] Quoted in Keynes, *The Life* (note 23), p. 388n. Today, peristalsis is understood as a progressive wave of contraction seen in tubes provided with longitudinal and transverse muscular fibres (e.g., the gastrointestinal tract). It consists in a narrowing and shortening of a portion of the tube, which then relaxes, while a distal portion becomes shortened and narrowed. By means of this motion the contents of the tube are forced toward the opening.

rely on images from vivisection or from anatomical drawing. Instead, he would require experimental analogues such as lute or bowstrings, for these, when stretched, may be set vibrating in different ways. But the lute and bow are important for another reason, for when a stretched string is plucked or otherwise acted upon, there is an interaction between two bodies: in the case of the lute, between string (or strings) and soundboard; in the case of the bow, between string and lath. Such analogues, therefore, might suggest an explanation of mind/body interaction as a conflict of forces. And the same analogues might suggest an explanation of tremor, which, physiologically, is a more or less regular rhythmic oscillation of part of a body around a fixed point or points involving alternate contraction of agonist and antagonist muscles.

Contrary to received opinion, Hobbes did utilise experimental analogues, as is evident from a number of his writings, including passages in *De corpore*.[27] Moreover, these analogues provided the model for his theory of motion and, especially, that part of it devoted to 'internal and invisible motions'. I begin, therefore, with a brief historical review of biomechanics, with special reference to forces that control body movement. But my main purpose is to elucidate Hobbes's chief dynamical construct, as it is found in *De corpore*, since this is the fundamental construct of his last philosophy.[28] Indeed, when properly understood, this construct explains Hobbes's physiology—organised body, as well as his pathology—disorganised body. In *De corpore* Hobbes had little to say about pathology, which, for him, was a study of the passions. Although he presented several such studies, I draw on the study presented in *Leviathan*, Hobbes's rhetorical masterpiece, which was written during the onset of

---

[27] For received opinion, see, e.g., R. Tuck, *Hobbes* (Oxford, 1989), pp. 49–50, who repeats, uncritically, the interpretation of S. Shapin and S. Schaffer, *Leviathan and the Air-Pump: Hobbes, Boyle, and the Experimental Life* (Princeton, 1985). The latter authors focus on Hobbes's criticisms of Boyle's vacuum experiments, so as to depict Hobbes as an anti-experimentalist. But Hobbes falls within the experimental tradition discussed by A. C. Crombie, 'Experimental science and the rational artist in early modern Europe', *Daedalus* (1986), 115: 49–74:

[28] I.e., works written after c.1636.

Parkinsonism and published in 1651.[29] Finally, I treat the reception of Hobbes's theory by some of his contemporaries, with particular attention to Robert Boyle, who sought to undermine the theory, and Robert Hooke, who promoted and developed the theory.

THE BIOMECHANICS OF *TONOI, TONOS* AND *TONIKE KINESIS*

From early times up to, and including, the seventeenth century, there was confusion about the cordlike substances in the body, which traditionally were called *neura* (from *neuein*: to bend) or *tenon* (from *tenein*: to stretch). In the Hippocratic corpus, however, the term *tonoi* is introduced and used interchangeably with *neura* to denote strong stretched cords, usually tendons or ligaments, that serve as a bond for the joints and give the body flexibility as well as the ability to bend and stretch.[30] Even though, in the same corpus, the heart is described as 'an exceedingly strong muscle—"muscle" in the sense not of "tendon" but of a compressed mass of flesh', there is no concept of muscle as an organ with a capacity for contracting and thereby performing work.[31]

Nor does Aristotle adumbrate the concept of muscle as an organ with a special function, even though he recognised muscles, but not nerves, as a separate tissue group.[32] Instead, he explained the mechanical aspect of movement by the tendons attached to the bones, whereby the joints are either pushed or pulled.[33] But, for him, tendons

---

[29] T. Hobbes, *Leviathan* [1651] ed. C. B. Macpherson (Harmondsworth, 1986). The sixteenth-century French physician, Jean Fernel, coined the terms 'physiology' and 'pathology'.

[30] For the nomenclature, see Hahm, *The Origins of Stoic Cosmology* (note 10), pp. 154–155, 168, 170–172, 179 n.45, and E. Bastholm, *The History of Muscle Physiology*...(Copenhagen, 1950), p. 21 *et passim*.

[31] G. E. R. Lloyd (ed.), *Hippocratic Writings* (Harmondsworth, 1986), p. 348.

[32] In Aristotle's technical vocabulary, *neura* denotes sinews, ligaments and muscles but not nerves. See F. Solmsen, 'Greek philosophy and the discovery of the nerves', *Museum Helveticum* (1961), 18: 150–167, 169–197 and 'Tissues and the soul: Philosophical contributions to physiology', *Philosophical Review* (1950), 59: 435–468.

[33] Aristotle, *De motu animalium* tr. by M. C. Nussbaum (Princeton, 1978); see also Bastholm, *The History of Muscle Physiology* (note 30), pp. 41–52.

and other bodily parts function in virtue of their connection to a motor centre, which Aristotle conceived as a fulcrum, the unmoving point from which opposite motions arise. Hence, functions immanent in organisation operate on the principle of the lever: just as the extremities of the balance arm move in a greater arc than do points closer to the centre, so too a small change at the centre of a symmetrical body initiates motion which becomes ample in the extremes.[34]

In Aristotle's biomechanics, there are three classes of animal movement—voluntary, involuntary and non-voluntary, the last being partly in our power, as when we decide to hold our breath for a short period of time. But a phantasm is involved in all three classes, the main difference being that involuntary and non-voluntary movements are initiated by appetite—phantasms without reasoning, whereas voluntary movements are initiated by desire—phantasms with reasoning. Hence, self-motion is possible only insofar as a living body has appetite; and soul, a force that Aristotle located in the heart, is the source of every vital change in, or movement of, a living body, although every such change involves both body and soul. But the instrument of soul, as well as the instrument of instruments, is heat—connate *pneuma*, so that the brain functions as a counterbalance by providing coolness.[35]

In their criticism of Aristotle, the Stoics retained the notion of the heart as a motor centre and seat of the principal part of an extended soul or warm breath—*pneuma*.[36] But the Stoics replaced Aristotle's lever principle with that of stretched strings to illustrate their doctrine that *pneuma* pervades and totally mixes with unformed

---

[34] For the importance of the lever principle in Aristotle's philosophy, see T. J. Tracy, *Physiological Theory and the Doctrine of the Mean in Plato and Aristotle* (Chicago, 1969).

[35] Aristotle, *De motu animalium* (note 33), seems to introduce connate *pneuma* as a development of his earlier doctrine of innate heat. See E. Mendelsohn, *Heat and Life: The Development of the Theory of Animal Heat* (Cambridge, Massachusetts, 1964), pp. 11–14, who remarks that vital heat, as defined by Aristotle, was utilised in explanations by the foremost physicians and physiologists well into the seventeenth century.

[36] See, e.g., B. Inwood, *Ethics and Human Action in Early Stoicism* (Oxford, 1985), pp. 1–17 *et passim*.

body so as to act as a cohesive and binding force—*tonos*.[37] In one important illustration, for example, Chrysippus compared the universe to a lyre that coheres and is harmonious, because it is bound together by *tonoi*.[38] Like lyre strings, *tonoi* function as a binding agent when they possess tension; and when they lack tension, the cosmic lyre, no longer bound up begins to dissolve. But *tonos* also causes cohesiveness in humans, for Chrysippus compared the *tonoi* of the body to the *tonoi* of the soul. The former are said to be well or ill tuned in respect to cordlike substances, whereas the latter may be spoken of as in or out of tune.[39]

The Stoics, therefore, developed a theory of the soul as an attunement. But their doctrine must be distinguished from the Platonic (Pythagorean) one, in which the soul is conceived as a harmony of abstract numbers and the relation of soul to body as an organisation by a pre-existing harmony of numbers that outlasts the body. Against this dualist theory, the Stoics conceived soul as a body—*pneuma*, thereby treating the relation of soul to body as an interaction or sympathy between two bodies, the cohesion of which is maintained by *tonos*. Hence, *pneuma* functions not only as a binding force but also as an instrument that generates physical states in body, the cordlike substances of which contract and expand with temperature changes. Such substances, therefore, require continual attunement, like strings of a lyre; and this attunement is affected by *tonike kinesis*—tonic motion.[40]

---

[37] For bow and lyre strings, see Hahm, *The Origin of Stoic Cosmology* (note 10), pp. 153–154; for taut threads of a spider's web, see S. Sambursky, *Physics of the Stoics* (London, 1987), pp. 24–25, 123–124. The pre-Socratic philosopher, Heraclitus, used all three as analogues of 'back-stretched connection' (i.e., tension).

[38] J. Rist, 'On Greek biology, Greek cosmology and some sources of theological pneuma', *Prudentia* (1985), supplementary number: 27–47, p. 45. Hahm, *The Origin of Stoic Cosmology* (note 10), p. 169, claims that Chrysippus gave up the imagery of the lyre, thereby failing to notice the change in emphasis in the Stoic imagery from that of a plucked lyre string (Cleanthes) to that of the whole lyre bound together by *tonoi* (Chrysippus).

[39] Ibid., pp. 169, 171; see also J. B. Gould, *The Philosophy of Chrysippus* (Leiden, 1970), p. 195.

[40] For the Stoic definition of 'primary impulse' as the power—*tonike dunamis*—to set oneself in motion in pursuit of goals, see Inwood, *Ethics and Action* (note

Doxographers do not agree on the meaning of this term,[41] despite the testimony of Galen, who conceived tonic motion as a local motion, when he identified four kinds of muscular motion: (1) the active motion of contraction; (2) the passive motion of relaxation due to the contraction of an antagonist muscle; (3) motion that involves neither contraction nor relaxation but which is the result of gravity—as when a raised arm falls; and (4) apparent immobility that still involves the active motion of contraction—as when the muscles of an extended arm under strain are in a state of excitation (which implies motion) and yet at rest as a whole.[42] It was this last kind of motion that Galen called '*tonike kinesis*' and which we now call 'active posture'.[43]

Galen thus identified contraction as the chief function of muscle; and he also recognised that muscles are innervated: nerves are resolved in the muscle, they assemble again into thicker branches and finally leave the muscle in the form of a tendon, to which is added ligaments and bands by which the muscle issues from and is attached to the bone. Nevertheless, he treated muscle as a kind of supporting tissue or packing between skin and bones, because he retained Aristotle's doctrine that tendons bend the joints. And Galen, too, believed that body is wholly in service of the soul, since muscle can never reach complete contraction without the cooperation of *tonos tes psyches*, a force that Galen seated in the brain and distributed to the various muscle groups via nerve channels.

Despite relocating the motor centre, Galen retained Aristotle's notion of the heart as a container by distinguishing between two types of fibre, voluntary and involuntary. The former, which run in the same lengthwise or crosswise direction, are muscle; and the latter, which run in several directions—as in the heart—are muscle-like.

---

36), pp. 27–28. Hence, the precondition for primary impulse is tautness—*syntonia*.

[41] See Hahm, *The Origin of Stoic Cosmology* (note 10), pp. 167, 182–183 n. 77.

[42] Bastholm, *The History of Muscle Physiology* (note 30), pp. 74–96; Sambursky, *Physics of the Stoics* (note 37), pp. 32–34.

[43] C. Sherrington, 'A note on the history of the word "tonus" as a physiological term', *Contributions to Medical and Biological Research*...ed. by C. L Dana (2 vols., New York, 1919), vol. 1, pp. 261–268; F. Fearing, 'The reflex maintenance of posture', *Reflex Action: A Study in the History of Physiological Psychology* (New York, 1964), pp. 218–231.

Nevertheless, he believed that an innate tension or *tonos symphytos* is present in all parts of the body, though variable in strength, since both types of fibre contract and, hence, perform work. But Galen referred the work of involuntary fibres to gross peristaltic motions of whole organs so as to explain the attraction of food, its assimilation and retention, and the discharge of surplus material, thereby restricting the work of voluntary fibres to the four muscular motions previously enumerated.

From the foregoing, we may identify three problems requiring solution. First, there is the nature of contractile substance: is it muscle or tendon? According to Galen, muscles consist of a fleshy part, serving as packing material, and a fibrous part that passes into and is continuous with the fibres of the tendon at each end, the tendon being the contractile substance. In the sixteenth century a small number of writers, including Vesalius, disputed the tendon theory by arguing that the fleshy part contracts.[44] But if muscle itself is the contractile substance, what is the contractile element: parenchyma or fibre? Galen's answer was: the ligamentous elements that spread into the muscle from the tendon. Yet, no definitive answer could be given to this question until the microscope revealed finer anatomical detail; and even then the notion of fibre as the element of form and function of the organism was slow to develop.[45]

Second, there is the characterisation of muscles as organs of voluntary movement, since, according to Aristotle and Galen, there are no mindless movements; and movement itself is not a primary expression of life, since it is a process emanating from the soul in which muscles have no share. Yet, there are many movements of

---

[44] O. Temkin, 'Vesalius on an immanent biological motor force', *Bulletin of the History of Medicine* (1965), 39: 277–280.

[45] In 1664 Niels Stensen argued that muscles do not consist of parenchyma and fibres; rather, they are collections of motor fibres that may be divided into minor fibrils bounded together by continuous transverse fibres of the membrane of the muscle. See E. Bastholm, 'Niels Stensen's myology', *Steno and Brain Research in the 17th Century* ed. by G. Scherz (Oxford, 1968), pp. 147–153. Long after the seventeenth century, nerves continued to be conceived not as bundles of fibres but as tubular prolongations of the substance of the central nervous system, although there were some early sceptics, including Vesalius and his pupil, Gabrielo Falloppio. See E. Clarke, 'The doctrine of the hollow nerve in the seventeenth and eighteenth centuries', *Medicine, Science and Culture*...ed. by L. G. Stevenson and R. P. Multhauf (Baltimore, 1968), pp. 123–141.

which we are aware but cannot control, as the Stoics seem to have recognised when they classed tremor from fear as a voluntary reaction and tremor from sudden shock as an involuntary reaction.[46] And much later, in the notebooks of Leonardo da Vinci, we find the observation that paralytics or those who are shivering or benumbed by cold move their trembling limbs such as the head or the hands without permission of the soul; which soul with all its powers cannot prevent these limbs from trembling. The same happens in the case of epilepsy or with severed limbs such as the tails of lizards.[47]

For a solution to this second problem, two things were required. First, it was necessary to redefine muscles as organs of movement. This characterisation is implicit in the writings of Jean Fernel, who pointed out that even the strongest will is unable to cause movement in a paralysed limb and offered this as proof that movement has relation not to appetite or desire but to the capacity of a muscle.[48] Second and related to the first, it was necessary to treat some muscular movements as mindless, that is to say, as reflex. In 1632 René Descartes wrote a tract devoted chiefly to mindless movements; and in 1664 Thomas Willis utilised the term 'reflection', when attempting to describe the neural processing of various reflex actions.[49]

Finally, there is a third problem—*tonike kinesis*, the motion underlying static equilibrium, as in active posture or tetanus. Galen supposed that the cause of static equilibrium was equal but opposing forces inside, not outside, the body, although the rapid succession of motions in opposite directions occur at such speed that, to the

---

[46] Inwood, *Ethics and Human Action* (note 36), p. 176. According to Aristotle, *De motu animalium* (note 35), p. 42,

> sense-perceptions are at once a kind of alteration[,] and *phantasiai* and thinking have the power of the actual things. For it turns out that the form conceived of the … pleasant or fearful is like the actual thing itself. That is why we shudder and are frightened just thinking of something.

[47] L. da Vinci, *The Notebooks* tr. by E. MacCurdy (2 vols., London, 1956), vol. 1, p. 107.

[48] Bastholm, *The History of Muscle Physiology* (note 30), p. 117.

[49] See R. Descartes, *Treatise on Man* tr. by T. S. Hall (Cambridge, Massachusetts, 1972), which was written in 1632 but not published until 1662; see also T. Willis, *Cerebri anatome: Cui accessit nervorum descriptio et usus* (London, 1664) and Chapter 6 *infra*.

observer, the body seems to be in a state of rest. Although Galen thus described *tonike kinesis* as an oscillatory motion, he believed that an explanation of such motion was to be sought from mechanics, not from muscle physiology. But mechanics from the time of Galen to the seventeenth century was Aristotelian, and the Aristotelian doctrine of contrary motions denied that oscillations are, or can be, continuous and regular. It was Galileo Galilei who, in 1638, refuted the Aristotelian doctrine by demonstrating, from his work on the pendulum, that opposite motions can be continuous and regular.[50]

In the 1620s, before Galileo published his refutation, William Harvey reviewed the classical, as well as some sixteenth-century sources in his draft treatise on animal movement.[51] Against the tendon theory of Aristotle and Galen, he maintained, with Vesalius, that contractile substance is muscle flesh and contractile element, fleshy fibre.[52] But, in this text, he failed to resolve the problem of the motor centre, as is evident from his explorations of the role of the heart and brain by means of various musical and other analogies.[53] Nevertheless, there is at least one hint that Harvey may have been dissatisfied with the doctrine that, for self-motion, body is wholly in service of the soul, for he queried what kind of movement scratching is.[54] Moreover, he did not repeat the classical definition of muscles as organs of voluntary movement; instead, he emphasised three things: all movements begin from contraction;[55] muscles 'pulsate' when

---

[50] See P. Ariotti, 'Aspects of the conception and development of the pendulum in the 17th century', *Archive for History of Exact Sciences* (1971-72), 8: 329–410.

[51] W. Harvey, *De motu locali animalium 1627* tr. by G. Whitteridge (Cambridge, 1959). The draft is more like a commonplace book than a treatise, as is clear from the way Harvey recorded the etymology of *neura* and *tenon,* pp. 68 n.1, 69 and n.4, 71 and n.7, 83, 97, 115, 117; repeated Aristotle's conception of fear (see above note 46), pp. 37, 99; and reviewed ideas about tremor, pp. 57, 83, 97, 99, 103, 121, 125, 141, 149.

[52] Ibid., pp. 5, 89, 117, in contradistinction to his anatomy teacher in Padua, Girolamo Fabrici (Hieronymus Fabricius ab Aquapendente), who retained the tendon theory.

[53] Ibid., pp. 111, 150. See also Kassler, *Inner Music* (note 3), pp. 36–38.

[54] Harvey, *De motu locali animalium* (note 51), p. 122: '*quod scalpere*'.

[55] Ibid., pp. 45, 103.

working;[56] and coordinated movement relies on many muscles working together in 'harmony and rhythm'.[57]

In considering the problem of tonic motion, Harvey provided a number of instances, mostly from Galen—for example, tetanus, an arm outstretched, a bird hovering in air, a fish swimming upstream. But he also included leaning and standing when awake, as well as sitting, standing, walking, retaining excrements, keeping the mouth closed when asleep.[58] In quoting from, or paraphrasing part of, Galen's *De motu musculorum*, Harvey wrote that muscles 'act and are not moved in tonic motion'.[59] But he also recorded, that tonic motion, when it waxes and wanes (*vacillat*), creates tremor.[60] And in a separate section on tonic motion, Harvey commonplaced Galen's definition that 'tonic motion occurs when both muscles act against each other and do not yield'.[61] But he then added: 'this is only one kind of tonic motion. The other kind is seen when a muscle is kept contracted as happens in an extension made by a single or coincident action'.[62]

Of the greatest importance, however, are three other brief additions to the Galenic doctrine of tonic motion. First, when paraphrasing Galen that tonic motion of the parts occurs 'when two

---

[56] Ibid., p. 111.

[57] Ibid., pp. 67, 111, 143, 145, 147, 149, 153.

[58] Ibid., pp. 27, 29, 61, 65, 83, 119, 121, which includes an instance of tonic motion 'in the French horse'. The translator glosses this as 'the instrument of torture known as chevalet'; but the context suggests one of the 'standing' positions taken by a horse in dressage. The terms for these positions were explained by William Cavendish, Hobbes's friend, in his treatise, *La méthode et invention nouvelle de dresser les chevaux* (Antwerp, 1658), for whom Hobbes wrote a paper entitled, 'Considerations touching the facility or difficulty of the motions of a horse on straight lines and circular'. See S. A. Strong, *A Catalogue of Letters and Other Historical Documents exhibited in the Library at Welbeck* (London, 1903); see also C. H. Firth (ed.), *The Life of William Cavendish, Duke of Newcastle...by Margaret, Duchess of Newcastle* (2d edn., London, n.d.), pp. xiv–xvi, xx–xxii, 63–63 nn.1–2, 106 and n.1, 206.

[59] Harvey, *De motu locali animalium* (note 51), p. 57.

[60] Ibid.; see also pp. 83, 121.

[61] Ibid., p. 119.

[62] Ibid.

muscles act in opposition to each other whose weakness gives tremor', Harvey added: 'balance needle'. Second, when surveying his teacher's version of the Galenic doctrine that holding is effected 'by having one part without movement or partly without movement as in tonic motion' and weakness is revealed as tremor, Harvey added: 'balance, compass'. And third, in considering perhaps his own notion of the pulsation of muscles 'according to rhythm', Harvey queried: 'Is alternation like iron in equilibrium? Or like the compass to the lodestone?'[63] All these additions suggest that Harvey conceived the motion that gives tone by analogy to the minute jiggle of the needle of a compass when excited by magnetism.[64] But there is an added implication that tremor appears when tonic motion lacks harmony and rhythm.

With this background in mind, we may now turn to Hobbes's own movement disorder, for the most disabling symptom of Parkinsonism is an impairment of the voluntary control of movement. It is this impairment that the patient experiences as an inner conflict between the forces of impulse and resistance or, to use psychological language, one's body won't do what one tells it to.[65] Sometimes the patient may experience the conflict as an obstructive will, when the performance of normal actions is rendered difficult or impossible; other times, as an explosive will, when dominant, abnormal actions are irresistible. In both types of experience, patients find 'that as soon as they "will" or intend or attempt a movement, a "counter-will" ... rises up to meet them'.[66] From the patient's point of view, therefore,

---

[63] Ibid., pp. 121, 141, 153.

[64] Perhaps from W. Gilbert, *De magnete* [1600] tr. by P. F. Mottelay (New York, 1958), pp. 192, 353. For Harvey's ideas about action at a distance, see Kassler, *Inner Music* (note 3), pp. 68–69, 240–241 n.92.

[65] For an extensive treatment of Parkinsonism from the physician's, as well as from the patient's point of view, see O. Sacks, *Awakenings* (London, 1982).

[66] Ibid., pp. 10, 38; cf. W. James, *Textbook of Psychology* (London, 1892), p. 454. The distress that results may call forth a variety of other symptoms, so there is need to alleviate distress by helping to make movement easy. One of the most effective ways of doing this, according to Sacks, ibid., pp. 56, 99, 115, 148, 283, 317, is through the power of music. Since the disorders of movement and force result in a loss of naturalness in posture and action, music supplies externally what the Parkinsonian patient lacks internally, for it functions as an

Parkinsonism is a conative disorder, that is, a disorder of effort or endeavour.[67]

HOBBES'S MECHANICS: ENDEAVOUR AS TENSION

The term, endeavour (*conatus*), is Hobbes's fundamental dynamical construct; and in *De corpore* he provided its only complete exposition. According to the developing argument presented there, endeavour is a process entailing deformation ('action') and restitution ('reaction') through continuous intervals to and through the infinitesimal vibratory motion that constitutes rest. To explicate this process, Hobbes had recourse to a number of different kinds of stretched strings, including the archer's monochord, as follows:

> when the lath of a cross-bow bent doth, as soon as it is at liberty, restore itself, though to him that judges by sense, both it and all its parts seem to be at rest; yet he, that judging by reason doth not account the taking away of impediment for an efficient cause, nor conceives that without an efficient cause any thing can pass from rest to motion, will conclude that the parts were already in motion before they began to restore themselves.[68]

All bodies, according to Hobbes, have 'the beginning of their restitution within themselves, namely, a certain motion in their internal parts, which was there, when, before the taking away of the force, they were compressed, or extended'.[69] Hobbes, therefore, conceived rest not as the opposite of motion but as thoroughly kinetic—like the indiscernible motion or *z* force that maintains tone.

In this neat example we have a demonstration of elasticity, as well as conservation. Elasticity relates to the possibility of getting mechanical work back from a system after it has been deformed by mechanical work done on it, whereas conservation relates to the

---

artificial pace-maker, thus restoring, temporarily, each patient's natural 'harmony and rhythm'.

[67] Ibid., p. 11: Parkinsonism 'exhibits a formal analogy of conative structure'.

[68] Hobbes, 'Concerning body' (note 7), pp. 347–348; for details, see Kassler, *Inner Music* (note 3), pp. 82–85 and sources cited.

[69] Hobbes, 'Concerning body' (note 7), p. 344.

possibility that upon deformation, energy is not lost but is transformed into the internal motion of the body's constituent particles. In practical cases, some of the mechanical energy put in is dissipated in overcoming internal frictional forces during the deformation. And some of the stored energy is dissipated against internal friction during the relaxation phase. Consequently, we cannot get back quite as much mechanical work as we put in, and the work 'lost' is dissipated as heat.[70] In ideal cases, the whole of the work done on the body during deformation can be recovered during relaxation. This implies not only that the body returns precisely to its original shape and dimensions when the deforming force is removed, but also that at each value of the deforming force the magnitude of the deformation is the same no matter whether the force is increasing or decreasing.

The study of deformable media is called 'continuum mechanics', the aim of which is to give theoretical insight into deformation and flow. In this kind of study, the kinematics deals mainly with motions that are deformations, whereas the dynamics deals mainly with stress, that is, with the forces exerted on one part of a body by a neighbouring part. The detailed discrete nature of the atomic and molecular structure is ignored, because it is assumed that the net effect of the atomic and molecular forces and motions can be adequately represented by the stress, the strain or rate of strain, and an appropriate constitutive relation. Although Hobbes introduced continuum mechanics to English readers, it was his younger contemporary, Robert Hooke, who provided the first constitutive relation known as Hooke's Law: *ut tensio, sic vis*—as the stretching, so the force.[71]

Nevertheless, Hobbes's example of the crossbow demonstrates the relation between stress and strain, for until the bow is strung, the

---

[70] For Hobbes's kinetic theory of heat, see ibid., pp. 324–325 *et passim*.

[71] *Tensio* (from *tendo*): to stretch, strain; to aim, shoot (an arrow); to bend (a bow); to tune (an instrument); to endeavour, exert oneself. Hooke related stress and strain to a stretched elastic spring, but his constitutive relation applies also to stretched strings and other linear systems. Although Hooke was aware of this relation as early as 1658, he announced it in an anagram in 1676 and provided details in 1678. In contemporary biomechanics Hooke's Law appears in models of postural mechanisms and, especially, that mechanism which determines the tone of a muscle—the stretch reflex.

string lies slack. As the bow is bent, its 'spring' or elastic force is felt. When the string is fitted to the bow, it then bears the elastic force and acquires tension. When the string is pulled, it resists the pull; and when the string is released, it tends to spring back to its equilibrium position. The same is true of the lute, with its bundles of strings. And, according to Hobbes, it is true also of particles of matter, for when urged against its neighbour by some exciting force, a particle may, by a sharp effort and sudden recoil, deliver up its motion in virtue of the elastic force exerted between the two particles.[72] The greater the elastic force, the more rapid the delivery and recoil, and vice versa. Elastic force, therefore, is like a sharp muscular effort and sudden recoil, for it is action and reaction 'in the same instant'.

The phrase, 'in the same instant', occurs frequently in *De corpore* and, like Hobbes's other key terms, has a number of different applications. When understood in the context of processes that are perfectly inverse, the term, regardless of its application, signifies that point when the conclusion of one movement is identical with the beginning of the opposite and conversely.[73] For example, in continuous processes, such as oscillatory systems, 'in the same instant' is that point when kinetic motion ('action') and dynamic tension ('reaction') conjoin, thereby producing the 'entire cause' or turning point, in which action is reversed. And in continuous processes, such as the flow of impulses that constitute our sensory life, the turning point is that instant when we decide to act or not to act. Hobbes called this point 'the beginning of voluntary motion', and it is like the Stoic apprehension or grasping, that moment when the hand contracts into a fist.[74]

---

[72] For Hobbes's notion of action as kinetic motion (change) and reaction as dynamic tension (cause of change), see Kassler, *Inner Music* (note 3), pp. 82–85. For historical background, see J. L. Russell, 'Action and reaction before Newton', *The British Journal for the History of Science* (1976), 9: 25–38, who states that Newton did not create a law of action and reaction *ex nihilo*: 'It was there already, although associated to a large extent with an obsolete Aristotelian theory of motion'.

[73] For the perfectly inverse process called 'analysis and synthesis', see W. Sacksteder, 'Hobbes' Logistica: Definition and commentary', *Philosophy Research Archives* (1982), 8: 55–94.

[74] The Stoics identified three states in the cognitive process—presentation, assent and apprehension, the first of which is like an open hand with the fingers

If we apply Hobbes's endeavour construct to the psychomotor domain, there are three aspects to note. First, rest is only an appearance, because body is maintained in a permanent state of tension by a real but infinitesimal vibratory motion, the *z* force necessary for the performance of all action. This is like muscular tonus, the slight resistance that normal relaxed muscle offers to passive movement. Second, when a body is set vibrating by an impact, blow or friction, the reaction is equal to the action. This is like muscular contraction, the active resistance of a muscle to an extending or compressing force. And third, the two processes, taken together, constitute Hobbes's 'entire cause', that which adjusts the motions of deformation, thereby enabling the restoration of equilibrium. This is like the muscle sense, the feeling of movement, weight, resistance and position which gives us not only an image of our body and its relation to surrounding objects in space but also an awareness of ourselves.[75]

Endeavour, therefore, is a condition of physiological equilibrium that, for Hobbes, is always associated with the presence of life. But equilibrium is not static, for internal tension is a constant infinitesimal vibratory motion that changes with time. Thus, any disturbance of this constancy, resulting in an excess of tension and leading to the attainment of goal and quiescence, can be said to exhibit the operation of a dynamic equilibrium. A recurrence of the same disturbance leads to the perception of cues, to increased tension and to appropriate action. In other words, repeated disequilibrium leads to learning by experience, which Hobbes exemplified by an

---

outstretched, the second, like the fingers contracted a little, and the third, like the hand closed entirely. In Hobbes's version, 'Concerning body' (note 7), p. 392, the incoming sensory impulses reach 'the last yielding part [the brain]; which by reaction [i.e., resistance], in the same instant, if the reaction be strong enough, makes a phantasm; and a phantasm being made, perception is made together with it'. For Hobbes, therefore, the stages are sense, phantasm and impulse, the last of which involves a contraction of the pia mater as the end link in a chain of events between the incoming stimulus and the outgoing final reaction (e.g., to act or not to act). Regarding the function of the pia mater, as Hobbes understood it, see Kassler, *Inner Music* (note 3), pp. 117–118.

[75] The old term for muscle sense, 'kinaesthesis', was coined in the nineteenth century by H. Charlton Bastian; the new term, 'proprioception', was coined in the early part of the twentieth century by Charles Sherrington.

aspiring musician learning to finger a musical instrument.[76] The mechanisms of equilibrium ensure that the sentient, so long as it remains alive, will tend toward optimal functioning; when optima are attained, so too is equilibrium. Indeed, for Hobbes, the underlying goal of living systems is self-optimisation.

But how is this goal to be achieved? In *De corpore* Hobbes provided a hint, when, in the concluding chapter on gravity, he made an analogy between the force of attraction and the sympathetic vibration of lute strings.[77] When stretched, a musical string becomes an extended object that can store energy in a series of discrete modes like musical pitches.[78] And if we pluck a second musical string that is in tune with the first, the latter will begin to sound in sympathy. It is possible in this way to transfer sound energy from one musical string to another, and pitches that take up energy in this way are said to resonate. Resonance frequency requires minimum force to maintain the vibrations, thereby overcoming gravitational, damping and stiffness forces that tend to diminish them, so that resonance is a periodic forcing function that not only supplies energy but also supplies it rhythmically with temporal consistency. But to do this, the strings must be tuned to the same tension, for if the strings are out of tune, a transfer of energy cannot take place.[79]

If we apply Hobbes's mechanics to a living organism, then the basic criteria for self-preservation and self-restoration recur as types of vibratory motion. Since the organism, like a lute when played

---

[76] Hobbes, 'Concerning body' (note 7), pp. 349–350.

[77] Ibid., pp. 527–528; see also pp. 499–500 on the causes of the 'concent of sounds'. Today, we would say that the magnetic field vibrates transversely about the direction of propagation, just like a stretched string, with the magnetic field corresponding to the displacement.

[78] If a musician sounds a pitch on a musical string, which is a linear system, the result is a superposition of overtones on the fundamental tone. For Hobbes's treatment of superposition, see Kassler, *Inner Music* (note 3), pp. 3, 111–112.

[79] I.e., interference will take place, and this can be heard as beats, which are periodic fluctuations of loudness produced by the superposition of tones of close, but not identical, frequencies. In some cases, however, sound will be interdicted altogether. See H. F. Cohen, *Quantifying Music: The Science of Music at the First Stage of the Scientific Revolution, 1580–1650* (Dordrecht, 1984), p. 110, who states that Mersenne was one of the first theorists to discuss the phenomenon of beats.

upon, has continually changing degrees of tension, resistance (cohesive force) is important in Hobbes's conception of function. Indeed, the development of a living organism takes place because of the resisting power of the central systems (vascular and nervous), which are in a state of static tension. Like a lute with its bundles of stretched strings, each part is in relation to every other, so that any change in static tension will be due to elastic stress or strain, and the restoring force will be instrumental in all autonomous activities, including voluntary movement.

From the foregoing, we may conclude that Hobbes did not rely on a simple mechanism that produces tension in response to extension. Rather, he conceived body as a system of units—bundles of strings of different lengths and tensions. He then defined work as a change in the configuration of such a system in opposition to the forces resisting it, and he treated energy as the capacity to do work. In passing through any cycle of changes of configuration, a physical system ('body') does the same quantity of external work that is done upon it, so that the energy derived from systems without is compensated for by an equal amount of energy communicated to external systems. When forces are in disequilibrium, action continues until equilibrium is reached. In this way physical systems are conserved. But rest, conceived as a dynamic equilibrium, presupposes a certain degree of cohesiveness of the system in question, although this cohesiveness may be studied without making any special assertions about the material of the system concerned.[80] Instead, the fundamental premisses will derive from Hobbes' endeavour construct.

The dynamic aspect of this construct is the elastic force we call 'tension'. In physiological terms, it is the healthy tautness that our muscles keep, even when not engaged in maintaining posture or executing movements. Since antiquity this tautness has frequently been likened to the tension of a tuned lyre or bowstring and spoken of as tonus. But pathologists did not, and still do not, limit the term in this way. Rather, they employ tonus in connection with all conceivable parts, thus giving rise to a large number of related terms,

---

[80] For Hobbes, as for the Stoics, coherence is the criterion. In the case of bundles of vibrating strings, the initial conditions are each string's modulus of elasticity and density.

as one may discover from consulting any medical dictionary.[81] If we disregard atonia, which, for Hobbes, would mean death—even a living death,[82] we may identify two main pathological states, of which there are many different degrees. One state is diminished tonus or hypotonia, which means not merely a weakening of contractile parts but also a loss of elasticity or a decrease of cohesiveness in general. The other state is increased tonus or hypertonia, which denotes a strengthening of contractile parts, a gain in elasticity and an increase of cohesiveness. The one state is a deficit; the other, an excess of endeavour.

In *De corpore* Hobbes pointed out that lack of cohesion may deprive us of intelligence, since 'we observe our own body, and find that by the indisposition of the eyes, the brain, the nerves, and the heart, that is, by obstructions, stupidity, and debility, we are deprived of light, so that a fitting disposition of the organs to receive impressions from without is ... a necessary part of the cause of light'.[83] But in *Leviathan* he supposed that defects of constitution, as well as of character, arose chiefly from three types of passions that are contrary to nature, namely, weak, indifferent and excessive passions. For 'to have weak Passions, is Dulnesse; and to have Passions indifferently for every thing, GIDDINESSE, and *Distraction*; and to have stronger, and more vehement Passions for any thing, than is ordinarily seen in others, is that which men call MADNESSE'.[84]

If we recur to Hobbes's stretched-string analogues, we may understand weak and excessive passions as too little or too much tone. In the former state, the self will be slack and motion, sluggish, so that reactivity will produce too little impulse. That this is Hobbes's

---

[81] E.g., atonic, atonic bladder, atonic ulcer, catatonic, cerebrotonic, dystonic, epitomic, hypertonic, hypotonic, hypotonic diplegia, isotonic, myotonic, myotonic acquisita, myotonic congenita, myotonic paradoxica, myotonic dystrophy, somatotonic, sympathicotonic, syntonic, tonic-clonic, tonic postural epilepsy, tonic seizure, tonic spasm, viscerotonic, etc.

[82] For Hobbes, as for the Stoics, activity and existence are one, so that tonus is the condition for life, and death is a slackening of tension, where living death would be a profound state of stupor, unconsciousness or arrested activity, as in a deep narcosis.

[83] Hobbes, 'Concerning body' (note 7), p. 78; see also pp. 392, 393, 397, 400.

[84] Hobbes, *Leviathan* (note 29), p. 139.

meaning is clear from his definition of 'DULNESSE, *Stupidity*' and the like, the names of which 'signifie slownesse of motion, or difficulty to be moved'.[85] But in the latter state, too much tone, the self will be so highly strung that reactivity will produce extravagant impulse, and the result will be 'strange and unusuall behaviour'.[86] Both types of passions exhibit lack of control, since weak passions may result in '*self-conceipt*', which is 'great *Dejection* of mind', whereas excessive passions may result in '*vaine-Glory*; which is commonly called Pride'. The 'violence, or continuance' of both 'maketh Madness', because in one state we are enslaved by causeless fears and anxieties and in the other state, held in bondage to anger and its excesses, rage and fury. And between these extremes, Hobbes wrote, there is a legion of other passions, so that 'if the Excesses be madness, there is no doubt but the Passions themselves, when they tend to Evill, are degrees of the same'.[87]

How, then, are Hobbes's indifferent passions to be understood? Since he defined them as distraction—a divided self—we may conjecture that his analogue was a false string, which is thicker in one part of its length than in another.[88] When plucked, the thicker part vibrates more slowly and, hence, sounds lower, whereas the more slender part vibrates more swiftly and, hence, sounds more acute. The unequal degrees of motion and rest in such a string would interfere with the motion of other strings to which it might have a relation, thus affecting resonance frequency or preventing sympathetic vibration altogether. In pathological terms, there would be a delay, as well as an exaggeration of reactions, so that many voluntary motions would

---

[85] Ibid., p. 135.

[86] Ibid., p. 142.

[87] Ibid., p. 140.

[88] As a viol and, possibly also, a lute player, Hobbes would have had direct experience of false strings, although he could have learned about them from two other sources: music treatises, including Mersenne's *Harmonie universelle* (note 12), and musicians employed by his patrons—the Cavendish family, their relations and neighbours. For some of the musicians, see, e.g., Hulse, 'Hardwick MS 29' (note 18); D. C. Price, *Patrons and Musicians of the English Renaissance* (Cambridge, 1981); and G. A. Philipps, 'John Wilbye's other patrons: The Cavendishes and their place in English musical life during the Renaissance', *Music Review* (1977), 38: 81–93.

be accompanied by tremor. And if the movement is complex, there would be defects of adjustment, thereby producing incoordination.

Like weak and excessive passions, tremor and incoordination are signs of impairment in the control of voluntary movements. It is important to note, therefore, that a taut string is a rudimentary servomechanism: the output of the string is made to control its operation with a view to not allowing the output at any time to exceed or to be less than a certain value. Hence, there is a physical limit, for if the tension of the string exceeds a certain value, it is likely to break, whereas if it is less than a certain value, there will be no reactivity.[89] Just as there is a limited scale of tones in music, so too there is a limited scale of tensions in health. In both cases, music and health, the scale will consist of different frequencies of periodic vibrations, and the limits of the scale will be determined by the nature of the instruments, some of which will be more, others less resilient.

If the scale is exceeded on the minus side, the level of tonus will be too low for control to be maintained. When this happens, there will be a tendency to flaccidity, diminished strength, an obstructed will. As the degrees of tension rise, so too does power, considered as rates of energy expenditure. As the scale continues beyond measure, our complex internal tensions will consist of a mixture of a very large number of frequencies, so that if the scale is exceeded on the plus side, the level of tonus will be too high for control to be maintained. When this happens, there will be a tendency to rigidity, increased strength, an explosive will. Hence, a person may become a fast-moving projectile, for example, by shooting out the arm in anger or by flinging about the entire body in rage.

For Hobbes, therefore, passions are tendencies conceived as different degrees of tensions; and these, in turn, may incite affective states, the impulses or feeling tones by which we become aware of ourselves and the external world. But this kind of awareness is pre-rational (*pathos*), not rational (*ethos*), because passions are transient

---

[89] For rupture as loss of continuity, see Hobbes, 'Concerning body' (note 7), pp. 475–476. This may happen in one of two ways, for either the separation may begin in the outermost superficies and proceed successively to the innermost parts thereof; or it may proceed in the convex superficies of the bowed part of a body and proceed to the concave superficies. Both types of rupture may be demonstrated by a lute or viol string, excessively stretched beyond its tuned tension or excessively pulled or pushed from its equilibrium position.

not coherent tensions. To become coherent, or in sympathy with nature, the passions need tuning by 'Method, Culture, or Instruction', for by these we 'bend' our minds to the pursuit of truth.[90] In so doing, we become aware of ourselves reflexively as the subject and object of impulse, for we learn to test the generalisations we make from experience. Thus, it is in our power to develop conscience or right reason, the cohesive force that enables us to live a virtuous life. But virtue is not a static state; rather, it is a continual effort to fine-tune the passions, since, according to Hobbes,

> there can be no such thing as perpetuall Tranquillity of mind, while we live here; because Life it selfe is but Motion, and can never be without Desire, nor without Feare, no more than without Sense. What kind of Felicity God hath ordained to them that devoutly honour him, a man shall no sooner know, than enjoy; being joyes, that now are ... incomprehensible ... .[91]

Many commentators have puzzled over Hobbes's remarks about God; and many, even those of his own day, have suspected him of atheism. Yet, Hobbes's remarks are neither puzzling nor impious, for we find their parallel in the book of Job, the key biblical text for those in crisis and from which Hobbes took the title, *Leviathan*.[92] In the biblical text God and Satan join in a wager to shake Job's faith and unseat his undoubted piety. God allows Satan to scourge Job but to spare his life. In the ordeal that follows Job never wavers from his belief *that God is*, though he fails to understand God's silence and indifference. Hence, Job seeks, indeed demands, accountability for his suffering; but God, who demands unconditional worship, does not

---

[90] Hobbes, *Leviathan* (note 29), p. 134; see also pp. 91, 95, where Hobbes attributed night and day dreams to the 'distemper'—lack of attunement—'of some of the inward parts of the Body'.

[91] Hobbes, *Leviathan* (note 29, pp. 129–130; see also p. 430, where Hobbes asserted: 'the nature of God is incomprehensible; that is to say, we understand nothing of *what he is*, but *only that he is*; and therefore the Attributes we give him, are ... our desire to honor him with such names as we conceive most honorable amongst our selves'.

[92] See Kassler, *Inner Music* (note 3), p. 51. The same biblical text supplied Hobbes with the title for his book on the Civil War, *Behemoth or the Long Parliament* [1682] ed. by F. Tönnies with an introduction by S. Holmes (Chicago, 1990).

render accounts. By accepting that God is beyond causality, Job is restored to his previous condition, for he now understands that good and evil rest solely on conscience, that human suffering is not a punishment of guilt, that 'Perfect love casteth out fear', even though 'No tremendous emotion can be had without fear'. Thus does Job expand his conception of the world that is sensed, for his scale of values includes recognition of a moral order. Although people cannot explain the source of this order, an affirmation of it is a basis for courage.

You may remember that there are two stages to Job's ordeal. In the first, Satan destroys Job's sons; but this does not undermine Job's faith, and God offers this as a proof of Job's perfect understanding. In response Satan remarks that a man will suffer anything as long as he himself is left intact. Thus follows the second stage of Job's ordeal, for his body is covered 'with sore boils from the sole of his foot unto his crown'. Job then curses the day of his birth and longs for death, but 'it cometh not'. In Hobbes's life there also were two main crises—the Civil War during which the sons of England were killed, and the fever, abscesses and pain during which Hobbes wished to end his life. But there was this difference, for, unlike Job, Hobbes would never be restored to his former condition. Indeed, for thirty-two years, he would have to endure a chronic and progressively disordering syndrome, one that would steadily diminish his ability to act in the world.

ENDEAVOUR AS PRESSURE *VERSUS* ENDEAVOUR AS TENSION

We are now in a position to summarise the chief features of Hobbes's mechanics as a theory of vibration. According to that theory, action takes place in a strictly continuous medium, in which the propagation of action takes time, and its velocity depends on the mechanical properties of the medium. Since Hobbes conceived bodies as highly deformable and mutually superpositional ('penetrable'), elasticity and compressibility are properties of every portion of the medium, however small, so that even if the medium is divided indefinitely, it still will be strictly continuous. Hence, force is all-pervading but continuously differentiated, like bundles of stretched strings of different lengths and tensions. On this approach, Hobbes could provide a mechanism for conservation by conceiving rebound as an elastic process equalising action and reaction.

Hobbes began to develop this theory from about 1636 as a reinterpretation of Stoic naturalism. About the same time Hobbes's older contemporaries, René Descartes and Pierre Gassendi, sought to protect a supernaturalistic ontology against the naturalism of the Renaissance.[93] To accomplish their aims, the supernaturalists and their followers restricted the kinds of explanatory analogy admitted into natural philosophy by arguing two things: first, that only those actions which produce, or tend to produce, locomotion were to be accepted as fundamental; second, that change of motion in bodies was to be explained only by communication from outside and not by any innate power or striving within bodies themselves.[94]

In conceiving how mechanical action might be transmitted from one point to another, the supernaturalists chose an explanation of action by impact and pressure, in which impenetrability was taken to be an ultimate property of real but indiscernibly small, individualised bodies. Hence, force manifests itself as a series of contiguous impacts, motion proceeds by jerks from point to point in a line or trajectory, and mechanics is defined as a theory of percussion. By excluding the possibility of striving in bodies, the supernaturalists were unable to provide a mechanism for the conservation of motion. Hence, they proceeded on the assumption that, upon impact, motion will be lost, because impenetrable bodies will stop, not rebound.[95]

Here, then, we have two competing mechanics—that of the continuum (superposition, infinite divisibility) and that of the discrete (impenetrability, indivisibility). But in criticising Hobbes, the supernaturalists focused chiefly on two aspects of his mechanics. First, on his premiss of more or less elasticity, organic as well as

---

[93] See K. Hutchison, 'Supernaturalism and the mechanical philosophy', *History of Science* (1983), 21: 297–333.

[94] See M. Hesse, *Forces and Fields: The Concept of Action at a Distance in the History of Physics* (New York, 1962).

[95] According to W. L. Scott, *The Conflict between Atomism and Conservation Theory 1644–1860* (London, 1970), p. 6, Descartes assumed that hard bodies rebound; but the criticism of his laws of impact led to a different conclusion, and this was maintained by most supernaturalists up to and including Newton. See Hobbes, 'Concerning body' (note 7), pp. 334–335 *et passim*, who rejected the metaphysic of impenetrability.

inorganic bodies will have a power of recovery, so that Hobbes's theory of restitution seemed to reinstate innate powers. Second, on his premiss that rebound is an elastic process equalising action and reaction, Hobbes's theory seemed to lend itself to a possibility observed in antiquity, when Cicero's Stoic mouthpiece, Balbus, observed that

> nature persists and coheres by its own power without any help from the gods. There is indeed inherent ... a kind of harmony or 'sympathy' as the Greeks call it. But the greater it is in its own right, the less need it be regarded as the work of some divine power.[96]

Hobbes' theory, therefore, seemed to identify the creating principles of the universe with God and, hence, to accept that the universe is governed by necessity and not by a God who transcends creation. It is not surprising, therefore, that a number of English supernaturalists devoted considerable energy to promoting and developing percussion mechanics, so as to ensure that Hobbes' mechanics did not succeed. But the history of science teaches us that there was only one mechanics before Newton;[97] and this teaching is tacitly adopted by Steven Shapin and Simon Schaffer in their widely acclaimed book published in 1985.[98] These authors proceed on the assumption that the hostility between Hobbes and one of his fiercest critics, Robert Boyle, arose because of different approaches to scientific method. From what has been said, I do not think the case is so simple, as will appear from a closer examination of Hobbes's theory.

According to that theory, nature ('body') is conceived as an indefinitely continuous scale of tensional forces filling space. Hence, Hobbes could argue that empty space is not empty; rather, it is filled

---

[96] Cicero, *De natura deorum* (2: 27–30). Hobbes, *Leviathan* (note 29), pp. 473, 414, 471–478, defined miracle as 'a work of God, (besides his operation by the way of Nature, ordained in the Creation,) done for the making manifest to his elect, the mission of an extraordinary Minister for their salvation'; but since 'Miracles now cease', he cautioned against the 'Imposture of Miracles' by false prophets.

[97] See, e.g., E. J. Dijksterhuis, *The Mechanization of the World Picture* tr. by C. Dikshoorn (London, 1961).

[98] Shapin and Schaffer, *Leviathan and the Air-Pump* (note 27).

with kinetic energy, an energy that may be likened to the zero-point jiggling of a stretched string caused by the mote of energy remaining in the system when nothing is left. From this potential energy or *z* force, nature creates something from apparent nothing. Hobbes thus laid down a challenge to religious orthodoxy by presenting a solution to the paradox, *ex nihilo nihil fit.*[99] Indeed, Boyle singled out this very solution, when he asserted that Hobbes had built 'several things in his philosophy upon the creation of the world, and an infinite power: and how a thing material can create matter, and have an infinite power, I confess I do not understand'.[100]

Boyle's comment appears in an appendix, issued with the second edition of his tract, *Spring of the Air* (1662), in which he provided the first complete statement of what now is commonly but not very justifiably referred to as 'Boyle's Law': that the pressures and expansions of the 'air' are in reciprocal proportion.[101] By the term 'pressure' Boyle denoted 'endeavour outwards', and he conceived this state as a tendency to and not as an actual motion of particles constituting a fluid. A spring, not a crossbow, is the appropriate

---

[99] According to Colie, *Paradoxia epidemica* (note 17), p. 223, *nihil* paradoxes were engaged in 'an operation at once imitative and blasphemous, at once sacred and profane, since the formal paradox, conventionally regarded as low, parodied at the same time as it imitated the divine act of Creation'.

[100] Boyle, *The Works* (note 3), vol. 1, p. 187; see also p. 194, where Boyle again expressed his dissatisfaction with 'that principle of Mr. Hobbes, though it be the fundamental one of his philosophy'. See J. R. Jacob, 'Boyle's atomism and the restoration assault on pagan naturalism', *Social Studies of Science* (1978), 8: 211–233, who identifies two groups of naturalists attacked by Boyle. The first group, consisting of Roman Catholic scholastic philosophers, adumbrated what Jacob calls 'an heretical naturalism' that may be and often is materialistic but never mechanistic: 'It is animistic and asserts that the world is governed by non-mechanical vital forces or spiritual agencies of divinity'. The second group believed in God and a divine order, but 'their God does not exist outside of and above nature', for the divine order is immanent in nature. Nevertheless, they were regarded as atheists 'not because they deny God's existence but simply because they are not theists, though because they worship a God immanent in nature they might also be called deists of sorts'. Certainly, Hobbes cannot be classed within the first group; but see Kassler, *Inner Music* (note 3), pp 49–50, 237 n.2, for difficulties in classing him with the second group.

[101] C. Webster, 'The discovery of Boyle's Law, and the concept of the elasticity of the air in the seventeenth century', *Archive for History of Exact Sciences* (1962–66), 2: 441–502.

analogue for Boyle's concept, since, if squeezed, a spring tries to recover itself and exerts a certain force. In this instance force is considered as a pressure; and you can measure the force by seeing how much it will press things. If you take a spring, and press it an inch, it takes perhaps a force of 1 lb.; and it will take a force of 2 lbs. to press it in another inch. Or again, if you pull it out an inch, it takes a force of 1 lb.; and if you pull it out another inch, it takes a force of 2 lbs. This is Boyle's endeavour concept, for he conceived force as producing pressure and being measured by pressure.

Such a conception is suitable for observations on bodies at rest, but it is not suitable for observations on bodies that are free to move. In the latter case we require a different conception of force, according to which every change of motion in direction or in speed must be the result of force and must be proportional to that force. This was the definition of Galileo, who, in 1638, considered forces as proportional to the motions, and the motions as proportional to the forces. When we wish to measure a force in this way, we do not press it against springs to see how much it will press them in. What we do is to cause it to act on bodies that are free to move and see what motions it will produce in them.

Galileo was concerned chiefly with the action of a body (e.g., a pendulum) under the accelerating or speed-quickening force due to gravity, the attractive force of which on any body is always proportional to the mass of that body. But he also investigated what happens when a heavy weight is hung at the end of a horizontal beam, thereby introducing the new science called 'strength of materials'. In the discourse relating to this new science, Galileo did not consider either the bending or the compression of fibres that takes place on the under side of the beam.[102] But Hobbes did just this, when he considered 'all flexion', that is, what happens in both convex and concave sides of a stressed structure.[103] And he developed Galileo's

---

[102] G. Galilei, *Dialogues concerning Two New Sciences* tr. by H. Crew and A. de Salvio (New York, 1954), pp. 115–116. According to the translators: 'The one fundamental error which is implicitly introduced into this proposition [Proposition I] and which is carried through the entire discussion of the Second Day consists in a failure to see that, in such a beam, there must be equilibrium between the forces of tension and compression over any cross-section.'

[103] Hobbes, 'Concerning body' (note 7), p. 343, did not repeat Galileo's mistake of supposing that in a stressed structure fibres are inextensible.

strength of materials in another respect, for he considered the action of a body (e.g., a crossbow) under the force due to elasticity.

But Hobbes also used elastic bodies for a different purpose, when he proposed that the sympathetic resonance of lute strings might explain the action of a magnetic body under the force due to attraction. In 1662 Boyle rejected Hobbes's continuum theory in favour of an emission theory, when he ascribed attraction to the impact of specially shaped particles of subtle matter, whose stream lines (magnetic 'effluvia') issue from the lodestone and pass through the pores of nearby bodies. Nevertheless, he admitted that 'Mr. Hobbes has another hypothesis ... but I know, that divers learned writers have absolutely rejected it, and not one such, that I have heard of, has approved it'.[104] Amongst such learned writers were Joseph Glanvill and Henry More, who explained all action at a distance by recourse to a matter-governing *anima mundi* or subtle matter. According to this explanation, the attraction between lodestones, as well as the sympathy between strings and other physical objects, takes place because a subtle matter conveys information by pushing impenetrable particles in a straight line.[105]

Glanvill and More, therefore, advocated a translation theory, according to which energy is transferred directly into the work of running the machine ('body'). The resonance theory of Hobbes is different, for it holds that actions not evidently accomplished by direct contact of objects are transmitted over long distances by the propagation of disturbances (oscillations, vibrations, undulations) through a continuous medium that is more or less elastic. This is a 'thermodynamic', not a machine theory, because energy passes through an intermediate stage of heat, the *z* force that Hobbes conceived as an infinitesimal vibratory motion. This kind of motion

---

[104] Boyle, *The Works* (note 3), vol. 1, pp. 223–224.

[105] See, e.g., J. Glanvill, *Scepsis scientifica: or, Confest Ignorance, the Way to Science; in an Essay of the Vanity of Dogmatizing, and Confident Opinion* [1665] ed. by J. Owen (London, 1885), and H. More, 'An antidote against atheism [1652]', *Philosophical Writings* ed. by F. I. Mackinnon (New York, 1969), in both of which there is an attack on Hobbes. Although Glanvill and More may be described as emanation theorists and Boyle as an emission theorist, the two types of theorists shared a common approach, as is pointed out by E. A. Burtt, *The Metaphysical Foundations of Modern Physical Science* (Garden City, New York, 1954), p. 193.

implies reversible change, in which each state differs infinitesimally from that preceding it. But the energy called 'heat' does not run the machine; rather, it integrates the activities of the organism ('body').

Let us now recall Hobbes's theory that bodies seemingly at rest still have *z* force, because their internal parts jiggle. According to this theory, 'body' is governed by the strict continuity of the *z* force, which is Hobbes' version of the Stoic *tonos*, the heat or active force that binds the universe into a dynamic whole. Hobbes attributed changes in degrees of elastic force to temperature changes, which, in turn, are due to the infinitesimal vibratory motion that produces more or less heat. Depending on the degree of its 'fervour', therefore, *z* force causes the mercury in the thermometer to rise and fall, just as more or less heat makes a taut string swell or shrink.

The implications of this approach are spelled out by Samuel Sambursky, who states that the Stoics had

> hit upon an important physical law which applies to closed systems that are not subject to any interference. All the forces acting in such a system are inner forces and their sum total vanishes. Any partial system, being open to influences from other parts, is imperfect in the sense that no such conservation law can be formulated for it and that changes occurring in it are the result of forces which under the prevailing circumstances are external and imposed from the outside.[106]

In Hobbes's philosophy, all functions are immanent in organisation (hence, his emphasis, in *De corpore*, on internal freedom from contradiction). But in supernaturalism only some functions are immanent, so that a transcendent God must intervene to keep the universe from running down. It seemed, therefore, that Hobbes had 'struck at the root of morality', a phrase utilised by Isaac Newton in a letter of apology to John Locke, when, during an illness, he (Newton) had taken Locke for a 'Hobbist'.[107]

Historians of seventeenth-century science have devoted considerable attention to Boyle's conception of elasticity as pressure but have paid little attention to the history of elasticity as tension.

---

[106] Sambursky, *Physics of the Stoics* (note 37), p. 114.

[107] Quoted in K. Dewhurst, *John Locke (1632–1704) Physician and Philosopher: A Medical Biography* (London, 1963), p. 285.

They also tend to confuse the forces called 'tension' and 'pressure', and the problem is compounded by inexact usage.[108] In physics, elastic force is defined as a constrained condition of the particles of a body when subjected to forces acting in opposite directions away from each other (usually along the body's greatest length), thus tending to draw them apart, balanced by forces of cohesion holding them together. This was Hobbes's definition. In common parlance, we refer to this force, or combination of forces, as tension. The same term is used inexactly to denote the expansive force of gas or vapour, properly called 'pressure'. This was Boyle's definition. To make the distinction precise, therefore, it was necessary to establish a constitutive relation for elasticity as tension. In 1674 Hobbes's friend, Petty, suggested duplicate proportion, that (in his terms) the ratio of the 'depressing powers' equals the ratio of the squares of the 'spaces of depression'.

This suggestion occurs in a tract that includes practical applications of 'measure, number and weight' to sound, pendulums, strings and bells, along with an hypothesis for magnetism, in which Petty conceived atoms as magnets.[109] But it was Petty's quantitative, not his atomistic, approach that Hobbes praised in a letter to Aubrey, written in February of the same year, where he stated:

> if I had seen his [Petty's] book before it went to the press I would not (as he thinks) have hindered it, but done as the [Royal] Society did, that is, urged him to print it. For the doctrine is easy to be demonstrated. The last Chapter which is of Elasticity is different from the Principle which I have taken for natural Philosophy; but I am of

---

[108] E.g., H. von Foerster, M. Mead, H. L. Teuber (eds.), *Cybernetics: Circular Causal and Feedback Mechanisms in Biological and Social System Transactions of the Ninth Conference...*(New York, 1953), pp. 6–47 and pp. xvi, 3, 51–52, 68, 70, 131, where the participants fail to recognise that biomedical and social sciences track scientific conceptual changes and technological concepts. Hence, tension, when applied metaphorically, say, to physiological psychology, may derive either from physical tension, as defined in this present chapter, or from electrical tension, defined before c.1882 as potential, electromotive or mechanical force exerted by electricity and after c.1882 as the stress along lines of a force in a dialectric.

[109] W. Petty, *A Discourse made before the Royal Society ... concerning ... Duplicate Proportion* (London, 1674).

> opinion that his supposition [about duplicate proportion] is very true, and will go a great way.[110]

Petty's mathematical formulation, which is not very clear, seems to relate to pressure ('depressing powers') and, thus, to bodies at rest. In 1677 Hooke offered a different version of duplicate proportion, one more appropriate to bodies in motion: 'that the proportion of the strength or power of moving any Body is always in a duplicate proportion of the Velocity it receives from it'. This relation, he stated, is a 'General Rule of Mechanicks' and is

> most certainly true in the motion of Bullets shot out of Cannons, Muskets, Pistols, Wind-guns, Cross-bows, Spitting Trunks, and the like; as likewise in the motion of Arrows shot with Bows or Ballistæ; of Stones thrown by the hand, or with Slings; of Pendulums moved by Gravity or Weights; of Musical Strings; of Springs, and all other vibrating Bodies; of the motion of Wheels, Flies, &c. drawn and turned by Weights or Springs; of the motion of Perpendicularly or Obliquely falling Bodies; and in a word, of all other Mechanical and local motions, allowance only being made for the impediment of the Air or other Fluid Medium, through which the Body is moved.[111]

Then, in 1678, Hooke announced the elastic law that bears his name.[112]

---

[110] Quoted in Q. Skinner, 'Thomas Hobbes and his disciples in France and England', *Comparative Studies in History and Society* (1966), 8: 153–167, p. 165; see also Malcolm, *The Correspondence* (note 1), vol. 2, p. 751. For Nicholas Mercator's undated explication of Petty's duplicate proportion, see J. Aubrey, *On Education* ed. by J. E. Stephens (London, 1972), pp. 106–108. Mercator also wrote treatises on other aspects of applied mathematics, including one on music, for which see R. Hooke, *The Diary* ed. by H. W. Robinson and W. Adams (London, 1935), pp. 242, 254, and W. Holder, *A Treatise of the Natural Grounds and Principles of Harmony* [1694] (New York, 1967), pp. 104–106 (whose 'Friend' is Hooke).

[111] For Hooke's General Rule of Mechanics, see R. T. Gunther, *Early Science in Oxford* (15 vols., Oxford, 1923–67), vol. 8, pp. 186–187. For an earlier statement (1669), resulting from Hooke's efforts to demonstrate elastic rebound, see Gunther, ibid., vol. 6, pp. 346–348.

[112] For Hooke's Law, see Gunther, ibid., vol. 8, pp. 331–356. M. Hesse, 'Hooke's vibration theory and the isochrony of springs', *Isis* (1966), 57: 433–441, outlines the development of Hooke's theory without spotting the influence

It is well known that Hooke formulated this law much earlier than 1678. But it is not well known that his musical experiments demonstrated that Hobbes's mechanics was correct: bodies are more or less elastic and consist of parts in continual vibration; there are continuous forces of tension in body; and action at a distance is due to these forces, as Hobbes had hinted when he compared the force of attraction to the sympathetic resonance of lute strings.[113] In developing this hint, Hooke laid down three suppositions about magnetism: first, that 'all magnetical Bodies have the constituent Parts of them of equal magnitude and equal Tone'; second, that 'the Motion or Tone of one magnetical Body is convey'd to that of another by means of a Dense Medium'; and third, that 'the motion of the Dense Medium is Circular and Vibrating'. From these three suppositions, he wrote, 'all the *Phœnomena* of Magneticks will be most evidently and clearly, even a priori deduced'.[114] He then conjectured that the planets are kept in their orbits by a force analogous to that of magnetism.[115]

Like Hobbes, Hooke modelled his representation of nature by recourse to different elastic bodies, including musical strings.[116] Once stretched to its tuned tension, it is impossible to make a musical string alter the pitch of a note without altering its elastic force. This is

---

of Hobbes, although she states clearly what other historians tend to gloss over, namely, that Boyle appeared 'to feel little obligation to ascribe his assistant's ideas to their true author until forced to do so'. For Hobbes's influence on Hooke, see Kassler, *Inner Music* (note 3), pp. 128–129 *et passim.*

[113] In a series of lectures delivered between 1828 and 1830, Michael Faraday reported that Hooke, in his *Micrographia* (1665), made, the 'first correct observations' about the transmission of sound through solid bodies (i.e., that sound was propagated swifter through wire than air) and that his results were confirmed in Berlin in 1788. Faraday's source for this statement was C. Wheatstone, 'On the transmission of musical sounds through solid linear conductors, and on their subsequent reciprocation [1830]', *The Scientific Papers*...(London, 1879), pp. 47–63. See J. C. Kassler, *The Science of Music in Britain, 1714–1830: A Catalogue of Writings, Lectures and Inventions* (2 vols., New York, 1979).

[114] R. Hooke, *The Posthumous Works* ed. by R. Waller [1705]...(New York, 1969), p. 364.

[115] Ibid., pp. 183–185.

[116] For Hooke's musical analogues, see Kassler, *Inner Music* (note 3), pp. 124–159.

because the string, being elastic, obeys Hooke's Law, that the time of back spring is invariable, so that the pitch of the note produced remains invariable, whatever the amplitude of vibration may be. Hooke recognised that upon this law depends the correct going of clocks and watches; and he speculated that upon this law also depends the elastic vibrations giving rise to heat, light, sound, magnetism and gravitation. It was Hooke, therefore, who provided the whole basis for a theory of elasticity and for the science that treats wave motion in deformable material media; but it was Hobbes who opened the way for Hooke's achievement.

In carrying forward Hobbes's research programme, Hooke developed a more profound understanding not only of elasticity but also of resonance.[117] Indeed, in a remarkable series of experiments on bells, he revealed that these instruments 'echo' their own tone. This insight, in turn, led him to conceive 'body' as an incessantly ringing bell, so that the whole universe 'chimes', that is, resonates.[118] It is noteworthy, therefore, that at least one supernaturalist, Henry More, wrote against Hooke's theory, whereby Hooke responded by pointing out that More's hypothesis of an *anima mundi* ('Hylarchick Spirit')

> tends to nothing but the discouraging industry from searching into, and finding out the true causes of the Phenomena of Nature: And incourages Ignorance and Superstition by perswading nothing more can be known, and that the Spirit will do what it pleases. For if all things be done by an Hylarchick Spirit, that is, I know not what, and

---

[117] L. Brodsley, C. Frank and J. W. Steeds, 'Prince Rupert's drops', *Notes and Records of the Royal Society* (1986), 41: 10–26, point out that Hooke also was the first to understand the origin of thermal stress.

[118] For Hooke's experiments with bells, see Kassler, *Inner Music* (note 3), pp. 143–147. Perhaps following Hooke, Leonard Euler conceived the sun as an incessantly ringing bell. See A. J. L. James, 'Thermodynamics and sources of solar heat, 1846–1862', *The British Journal for the History of Science* (1982), 15: 155–181, who claims, p. 158, that Julius Mayer rejected the bell analogy. But see J. R. Mayer, 'Celestial dynamics', *Applications of Energy: Nineteenth Century* ed. by R. B. Lindsay (Stroudsburg, Pennsylvania, 1976), pp. 151–286, who, after comparing sound and light, observes, p. 152, that the 'sun has often and appropriately been compared to an incessantly sounding bell'. The analogy explained the radiation of heat but not its source, as Hooke grasped before Mayer when studying the source of earth tremors from volcanic eruptions. The former, but not the latter, can be explained by the bell analogy, just as, today, we describe the earth, during tremors, as 'ringing' like a bell.

> to be found I know not when or where, and acts all things I know not how, what should ... I trouble my self to enquire into that which is never to be understood, and is beyond the reach of my Faculties to comprehend?[119]

Hooke's more immediate audience, however, were Fellows of the Royal Society, where he performed many experiments to demonstrate elasticity, modes of vibration and resonance. But his ideas seemed to fall on deaf ears.[120] Or did they? That such might not have been the case appears from a tract with the title, *An Essay of the Great Effects of Even Languid and Unheeded Motion.* Published anonymously in 1685, the tract was reissued in the same year with Boyle's name and republished in 1690.[121] Ostensibly intended as part of Boyle's history of air, the *Essay* includes reports of experiments on sound generators and on the transmission of many different kinds of sounds—for example, the human voice, the tones of stretched strings, the ringing of bells, the report of ordnance, the noise of earthquakes and the clap of thunder, all of which were Hooke's special studies. According to an advertisement at the front of the tract, the portion relating to acoustics had been printed seven years earlier in 1679, the year of Hobbes's death.

It is highly probable, therefore, that the *Essay* constitutes a covert attack on the mechanics of Hobbes and Hooke; and this probability increases if we consider the following five points: First, in the two chapters relating to effluvia, Boyle reiterated an emission theory of magnetism.[122] Second, in the remaining chapters on acoustics, he included many of Hooke's experiments, though he failed to mention Hooke or to describe the experiments accurately (as

---

[119] For Hooke's answer to More, see Gunther, *Early Science in Oxford* (note 111), vol. 8, p. 188.

[120] The case was similar with Hooke's combustion theory, for which see D. McKie, 'Fire and the flamma vitalis: Boyle, Hooke and Mayow', *Science, Medicine and History*...ed. by E. A. Underwood (2 vols., London, 1953), vol. 1, pp. 469–488.

[121] Boyle, *The Works* (note 3), vol. 5, pp. 1–37.

[122] Ibid., p. 22.

will appear from the next three points).[123] Third, although Boyle allowed that solids might be more or less elastic (a real concession, since this would contradict the metaphysic of impenetrability), he suppressed the fact that different kinds of solids vibrate in different ways, as Hooke repeatedly demonstrated from 1665 onwards.[124] Fourth, in treating sound generation and transmission, Boyle conceived vibration as a random 'agitation' of particles and not as a simple harmonic motion, the importance of which Hooke understood so well, even though his 1678 analysis—one of the earliest in an elastic context, was flawed (because based on his General Rule of Mechanics).[125] Fifth and finally, although Boyle agreed that many effects at a distance are due to 'sympathy', he retained a translation theory in opposition to the resonance theory of Hobbes and Hooke.[126]

In the *Essay* Boyle focused on 'sympathy' and, hence, the problem of action at a distance, whereas in 1662 he was concerned chiefly with Hobbes's conception of a *z* force, according to which rest is a real but infinitesimal vibratory motion. Since this motion maintains body in a permanent state of tension, Hobbes propounded a dynamic conception of 'body', just as Harvey did, when he hinted that the indiscernible motion that gives tone is like the jiggle of the needle of a compass when excited by magnetism. It is tempting to believe that Hobbes recalled Harvey's insight when developing his conception of *z* force from experiments with stretched strings. Whether or not this was the case (for there is little evidence to prove

---

[123] Ibid., p. 4, where Boyle mentions Hooke only once, when reporting the theory of an 'ingenious person' on the cause of a flint's sparking (a marginal note identifies the person as Hooke).

[124] E.g., in describing Hooke's experiments with glass bells, Boyle, ibid., pp. 30–31, ignored the patterns of vibration of the contained substance, reporting only that the substance is agitated. By contrast Hooke's glass bell experiments provided a visible demonstration of patterns of vibration, a type of experiment Hooke devised as early as 1665 with sand on a stretched drum membrane. See Kassler, *Inner Music* (note 3), pp. 157–158.

[125] See, e.g., R. Westfall, *Force in Newton's Physics: The Science of Dynamics in the Seventeenth Century* (London, 1971), pp. 211–212, 260–261.

[126] Boyle's objections to the resonance theory of sympathy seem to have been made on the grounds that the motion of the untouched string had been caused not by the 'impulse' of the air but by the propagated motion of the instrument itself to which the touched string was also fastened.

it), Hobbes's experiments would have demonstrated that regular (i.e., periodic) vibratory motion produces 'tone' (pitch), but when that motion waxes and wanes, 'tremor' (beating) results.[127]

CONCLUSION

Although a number of physicians may be numbered among Hobbes's friends, biographers have yet to discover the name of the physician who treated Hobbes after his return to England from France. Hence, Aubrey remains our best source for Hobbes's regimen, which, for thirty-two years, emphasised 'moderation and regularity' thus: After waking at seven, Hobbes would breakfast, then walk and meditate till ten, when he would return to his rooms to 'putt downe the minutes of his thoughts'. At eleven, he would eat the main meal of the day, finishing off with a pipe of tobacco and a thirty- minute nap. In the afternoon he would write out the thoughts minuted in the morning; and before retiring, he would close his doors fast, so that no one could hear him. Then, when abed, he would pick up his 'bookes of prick-song lyeing on his table' and would sing 'aloud (not that he had a very good voice) but for his health's Sake: [for] he did beleeve it did his Lunges good, and conduced much to prolong his life'.[128]

---

[127] A periodic vibration is heard as a tone; for beats, see above (note 79).

[128] Aubrey, *Brief Lives* (note 2), p. 315. For historical background relating to singing for health, see G. Finney, 'Medical theories of vocal exercise and health', *Bulletin of the History of Medicine* (1968), 40: 422–449.

6. BLOWING THE ORGAN
WILLIS, HYDRO-PNEUMATICS AND HIERARCHY

We think playing upon man is like playing upon an ordinary organ.
It is indeed an organ, but strange, shifting and changeable.
Those who only know how to play an ordinary organ
would never be in tune on this one.
You have to know where the keys are.
B. Pascal, *Pensées* (c.1659, III.111)

INTRODUCTION

From the time of the ancient Greeks, there has been a recurring theme in which the passions are conceived as storms at sea or tempests in the air. In the seventeenth century this theme occurs in two different versions. One version, deriving from the Stoics, regards passions as fatal to a tranquil mind, so that Francis Bacon, for example, wrote: 'as the sea would of itself be calm and quiet, if the winds did not move and trouble it; so ... the mind ... would be temperate and stayed, if the affections, as winds, did not put it into tumult and perturbation'.[1] The other version, deriving from Aristotle, regards the passions as spurs to action, so that Henry More, for example, maintained that the brisk winds of passion serve as active forces guarding man's vessel from inertia.[2] To extend the metaphor of the passions as winds, one might conceive the body as a ship, either moved by healthy gales or tossed hither and thither by every wind, and the mind as a pilot or steersman,

[1] F. Bacon, *The Advancement of Learning* [1605]...ed. by A. Johnston (Oxford, 1974), p. 163.

[2] H. More, *Enchiridion ethicum* [1667]...(New York, 1930), p. 34.

since reason is supposed to rule the passions.[3] But John Bramhall lamented that reason is too seldom the guide at the helm, because passion, 'like an unruly passenger ... thrusts reason away from the rudder'.[4]

Although the man of passion is commonly represented as sailing 'in the frail bark of the flesh through the waves of the world',[5] this chapter analyses a different representation, one that was utilised by Thomas Willis (1621–75), perhaps the most prominent seventeenth-century English physician after Harvey.[6] From the single contemporary source that describes Willis's medical practice, we learn that he was a caring physician;[7] and from his own writings we discover an astute anatomical and clinical observer.[8] But his philosophy is Janus-faced.[9] Anatomically, Willis put the mechanisms of the brain and nervous system on their modern footing, insofar as could then be done. But he eschewed the machine theory of Cartesianism, interpreted not only as pure mechanism but also as a

---

[3] See K. M. Grange, 'The ship symbol as a key to former theories of the emotions', *Bulletin of the History of Medicine* (1962), 36: 512–523.

[4] J. Bramhall, *The Works*...(5 vols., Oxford, 1842–45), vol. 4, p. 416.

[5] F. Bacon, 'Of adversity', *Essays* [1625]...(London, 1907), p. 27; for the tradition, see Grange, 'The ship symbol' (note 3).

[6] For biographical details, see H. Isler, *Thomas Willis 1621–1675: Doctor and Scientist* (New York, 1968); K. Dewhurst, *Thomas Willis as a Physician* (Los Angeles, 1964) and K. Dewhurst (ed.), *Willis's Oxford Casebook (1650–52)* (Oxford, 1981), pp. 1–40.

[7] The source is Roger North; see J. C. Kassler, *Inner Music: Hobbes, Hooke and North on Internal Character* (London, 1995), p. 263.

[8] See especially P. F. Cranefield, 'A seventeenth century view of mental deficiency and schizophrenia: Thomas Willis on "Stupidity or Foolishness"', *Bulletin of the History of Medicine* (1961), 35: 291–316; and W. S. Miller, 'Thomas Willis and his De phthisi pulmonari', *The American Review of Tuberculosis* (1922), 5: 934–949.

[9] My interpretation is based on T. Willis, *Cerebri anatome* (1664), tr. by S. Pordage in 1681 and partially reprinted in Volume 2 of W. Feindel (ed.), *The Anatomy of the Brain and Nerves*...(2 vols., Montreal, 1965); and *De anima brutorum* (1670), tr. by S. Pordage as *Two Discourses concerning the Soul of Brutes*...(London, 1683). Hereafter, these works are cited as Willis 1664/81 and Willis 1670/83.

dualism that seemed to have been carried too far, thereby posing the ultimate theological problem of God's power and presence in His creation.[10] To combat Cartesianism, therefore, Willis chose to re-animate the machine by recourse to spirits and other ancient doctrines deriving from the Hippocratics, Plato, Aristotle, Galen and others. In this way he hoped to harmonise his philosophy with his Anglican faith.[11]

In the course of his writings, Willis modified details of his philosophy, but two features remained constant. First, he conceived body as an hydraulic organ, inspired by gentle winds of instinct or vehement winds of passion. Those winds, arising automatically with reference to the welfare of the organism, result in the affections, conceived as a complex of internal motions—harmonious or disharmonious, the external signs of which enable the physician to determine which internal parts of the organism are well- or ill-tempered. Second, since hydraulic organs are compound machines that operate on an hydro-pneumatic principle, Willis could describe the person in meteorological terms: the normative, regenerated person is 'like a Calm Sea, with a smooth Superficies', whereas the unregenerate human equivalent is 'troubled', 'like water shaken into various Circles, and wavings by the blasts of the Winds, or by some solid thing cast into it'.[12]

Of Willis's several writings, two treatises are important for our theme. The first, published in 1664 as *Cerebri anatome*, became a foundation document in the anatomy of the brain and nervous system and was required reading for neuroanatomists until nearly the middle of the nineteenth century.[13] The second, published in 1672 as *De anima brutorum*, is a treatise on physiological psychology, in which

---

[10] Willis 1670/83, p. 3, accused Descartes of introducing a version of the Epicurean hypothesis when treating spirits as the essence of the 'Soul of Brutes'. For a brief treatment of the English response to Epicureanism, see H. Jones, *The Epicurean Tradition* (London, 1992), pp. 186–213.

[11] According to Dewhurst (ed.), *Willis's Oxford Casebook* (note 6), p. 6, the Civil War interrupted Willis's plans to become an Anglican clergyman.

[12] Willis 1670/83, p. 45.

[13] See R. G. Frank Jr., 'Thomas Willis', *Dictionary of Scientific Biography* ed. by C. C. Gillespie (16 vols., New York, 1970-80), vol. 14, pp. 404–409.

Willis repeated and improved upon some of the doctrines contained in the earlier work.[14] In the latter treatise Willis also provided the most overt clue to his model, when he compared humans and animals to different kinds of hydraulic organs. Now, in all kinds of hydraulic organs, there are four common features: (1) a wind-raising device, operated by lever, pulley or some other type of mechanism, directing air under pressure to (2) a chest in which wind is stored and admitted by (3) a mechanism operated by some kind of keyboard to (4) one or more rows of pipes. How, then, does the animal *hydraulis* differ from the human one?

An answer to this question may be gleaned from Willis's statement that

> although the Musical Organ very much requires the labour of him playing on it, by whose direction, the spirit or wind being admitted, now into these, anon into those, and into other Pipes, causes the manifold harmony, and almost infinite Varieties of Tunes; yet sometimes I have seen such an Instrument so prepared, that without any Musitian directing, the little doors being shut up, by a certain law and order, by the mere Course of a Water, almost the same harmony is made, and the same tunes equal with those Composed by Art.[15]

On the premiss that animals are pre- rather than self-determined, we may infer that their structure is like a water organ consisting of (1) a bellows and reservoir complete with operating mechanism, (2) a pallet chest combining soundboard and pipe rack with pipes, (3) a barrel, and (4) a key-frame and action parts. The first of the four parts would correspond to the animal's circulatory and respiratory systems, whereas the remaining parts would correspond to its nervous system, in which the barrel is the encephalon and the pallet, pipes and key-frame are its 'speaking' mechanism. According to this structural hypothesis, animals are automata, since, like clock organs, they will play only those tunes that are programmed on their barrel.

---

[14] The *OED* main and supplementary volumes give only three seventeenth-century sources for the term 'psychology' (1653, 1680, 1693), none of which are from Willis 1670/83, pp. 7 and 18.

[15] Willis 1670/83, pp. 33–34.

In his earlier, 1664 treatise Willis had already noted that animals are 'like Machines framed with a more simple furniture and with less workmanship, and therefore furnished with a motion of one kind only, or determined for the doing still the same thing', whereas in humans, 'divers series of motions, and as it were complications of wheels within wheels, appear'.[16] By the phrase, 'wheels within wheels', Willis denoted the fluids that circulate within the body, whereas by the term, 'motions', he signified the solid containers or mechanisms of the body. In what follows I shall outline the anatomy of the human *hydraulis* by considering its operating and control systems. I shall then treat its physiology by considering how the human *hydraulis* plays and is played upon, as well as some of the music it plays.

ORGAN STRUCTURE

In *Cerebri anatome* Willis developed in detail his model of the human *hydraulis* with its four main parts: first, air feeders and reservoir, complete with operating mechanisms; second, windchest, combining soundboard and pipe rack with pipes; third, a barrel pricked with automatic tunes; and fourth, a key-frame and action parts. The first of the four parts constitutes the operating system with its hydraulic and blowing mechanisms, whereas the remaining three parts constitute the control system with its key-frame, barrel and windchest mechanisms. The 'seats' of these two systems are, respectively, the precordia and the encephalon. In Willis's anatomy, as illustrated in *Figures 6.1–3*, the term 'precordia' includes the heart itself, as well as the parts about the heart—the lungs, pericardium, diaphragm or floor of the thorax, and the region below called 'hypochondria', whereas the term 'encephalon' signifies the cerebrum, cerebellum and their common appendix, the 'medullar Trunk'. Let us begin with the operating system, for this is the source of wind supply.

---

[16] Willis 1664/81, p. 162 and unpaginated table of hard words, 'Rotation, *A going round, like a wheel, a moving round*'.

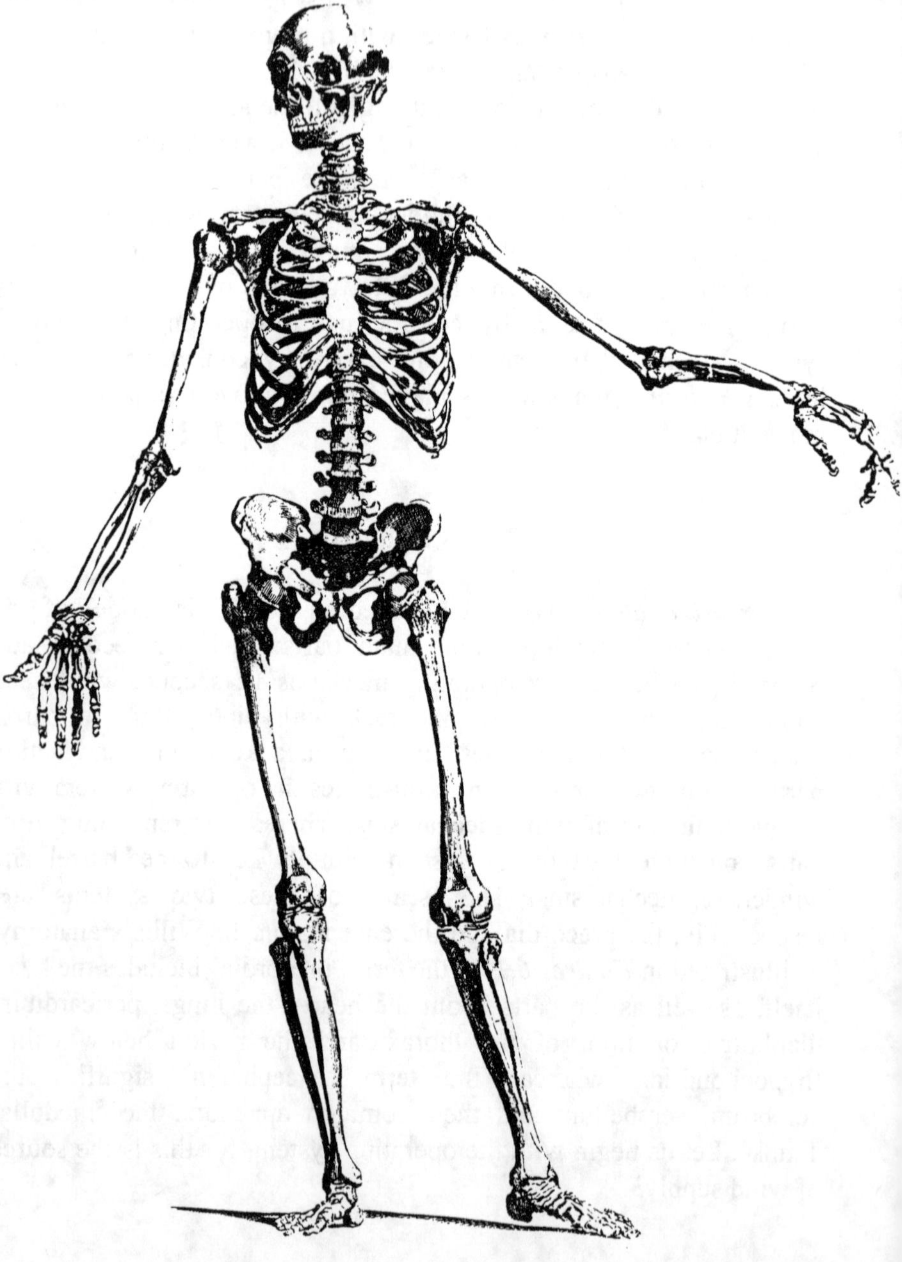

Figure 6.1. Frame of the human *hydraulis*

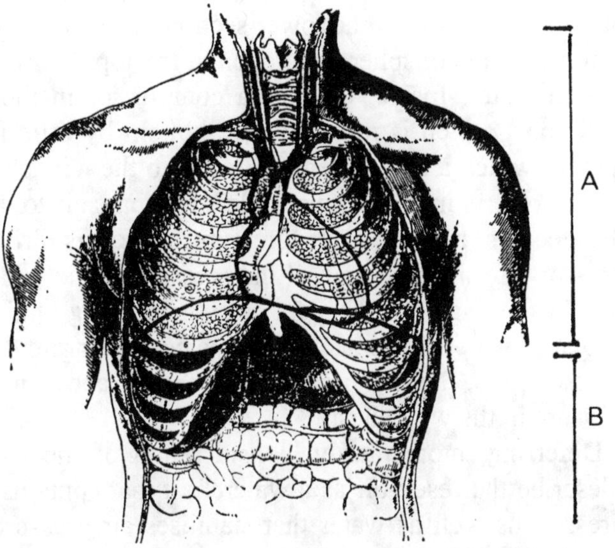

Figure 6.2. Precordia: seat of operating system, showing
(A) thoracic region with lungs ≈ air feeders and
heart enveloped by pericardium ≈ compressor in waterchest
(B) hypochondriac region (cut off)

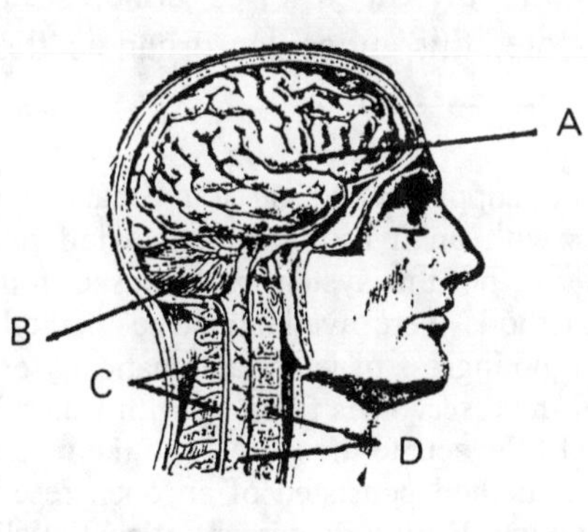

Figure 6.3. Encephalon: seat of control system, showing
(A) cerebrum ≈ keyboard mechanism,
(B) cerebellum ≈ barrel mechanism,
(C) medullar trunk (cut off) ≈ windchest with soundboard and
pipe rack (nerves ≈ pipes not shown)

In hydraulic organs two essential mechanisms constitute the operating system. These mechanisms are air feeders and a reservoir, the latter of which forms the lower section of the *hydraulis*.[17] Shaped in the form of a hemisphere and open at the top, the reservoir is half filled with water. Inside, there is a compressor in the shape of an inverted cone, which is raised off the bottom by little feet and from the apex of which a tube ascends directly to the windchest. On either side of the reservoir are two air feeders that pass air to the windchest. As the feeders raise the air pressure, water is driven from the compressor into the reservoir. If the pressure in the windchest drops, owing to a slowing of the feeders or to the flowing of air in the pipes, the displaced water flows back from the container and the level in the compressor rises, thereby compressing the air above it and restoring the pressure in the windchest.

Deferring, momentarily, consideration of the air feeders, we may describe the reservoir as a waterchest that contains an hydraulic compressor, as well as water that stabilises air pressure. In Willis's human *hydraulis* the waterchest is the pericardium, a closed membranous sac containing a clear serous liquid.[18] Situated in the middle part of the thorax between the two lobes of the lungs, the pericardium also envelops and fastens the heart, a muscular organ in the shape of an inverted cone (see *Figure 6.3*). From the heart all the veins terminate and all the arteries arise, so that by its alternate systole and diastole, this muscular organ is the principal force, or

---

[17] In antiquity wind supply relied on a mechanism of alternating pistons, employed together with an hydro-pneumatic wind reservoir. There is some doubt as to whether or not this system was utilised in the seventeenth century, when two other methods were available. One method consisted of a closed reservoir, in which the ingress of water automatically expels the air through an air outlet pipe until the reservoir is filled up with water. When so filled, a water outlet tap is opened to evacuate the water, and the reservoir once more fills up with air. The other method consisted of an open reservoir, into which water descends through a pipe. During its descent, it continually sucks down with it a volume of air, and this air-water mixture spurts into the reservoir. Since there is no closing tap in the lower water outlet pipe, there will be a constant stream of water spilling out from it. See J. J. L. Haspels, *Automatic Musical Instruments, their Mechanics and their Music 1580–1820* (Nirota, 1987).

[18] Willis 1664/81, p. 151 and unpaginated table, 'Pericardium, *The thin skin or membrane covering the whole heart like a case*'.

efficient instrument, of circulation.[19] Pumped by the heart, the blood flows through the blood vessels, slower or faster, depending on the force of propulsion, the density of fluids and the size, convolutions and resistance of the vessels. On contraction the blood begins its primary ('exterior') circulation through the arteries to all the parts of the body, returning through the veins to the heart. From thence the blood makes its secondary ('interior') circulation through the pulmonary artery to the lungs and back again to the heart by way of the pulmonary vein.[20]

In Willis's day two types of compressor could serve as analogues for the heart. One type is the piston pump, which is a rigid pump cylinder, either single or double acting.[21] But we may rule out this type, because the heart is a flexible, not a rigid organ. This leaves the second type of compressor, the bellows, which in its simplest form consists of a combination of bag and box, formed of an upper and lower board, joined by flexible leather sides, enclosing a cavity capable of contraction and expansion, and furnished with a valve opening inwards through which air enters and fills the expanded cavity and a tube or nozzle through which air is forced out in a stream when the instrument is compressed. It has many modifications of form and structure, according to its purpose, so that, for example, there are wind bellows that maintain air pressure, water bellows that maintain fluid pressure and blast bellows that maintain temperature in a blast furnace.

Since Willis conceived the heart's interior as a hearth ('fireplace') in which blood is 'boiled up', we may conjecture that his analogue is a pair of blast bellows, illustrated in *Figure 6.4*, in which a continuous blast is secured by one of the pair blowing and the other

---

[19] Ibid., p. 152.

[20] Ibid., pp. 155–156.

[21] In its simplest form the piston pump consisted of a barrel of wood or metal standing erect in the water, with a foot valve. Within the cylinder, there was a piston composed of two plates of copper with several layers of leather between them. As the piston was raised, the cylinder was filled by suction; on the downstroke the foot valve closed and the water was forced through the outlet, a check valve being set in the delivery pipe to stop water returning into the pump.

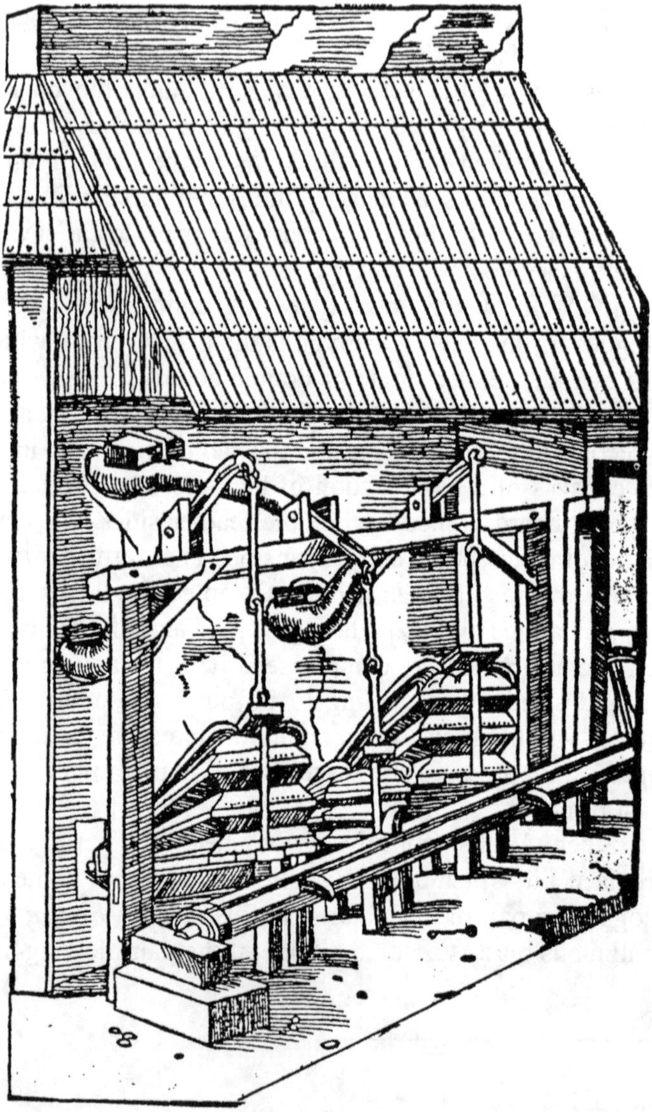

Figure 6.4. Back wall of blast furnace, showing holes lined with two nozzles each connected to a pair of blast bellows

filling.[22] Our conjecture is commensurate with the structure of the heart itself, the two halves of which pump in series: the right half feeds spent blood fluid out to the lungs to return to the larger pump at the left side, which, in turn, sends out aerated and more crimson blood to bathe nearly every cell of the body, the outflow rejoining eventually to the right side of the heart.[23] But our conjecture also is commensurate with Willis's belief that a chemical process in the heart enkindles the blood, so that it regains the heat for its journey by the arteries that it lost when returning to the heart by the veins.[24] During enkindling the blood 'boils up', releasing heat as vapour; and this vapour 'blows up' the heart, which increases in volume, just as a bellows seems to increase its volume upon inflation.[25] Willis, therefore, equated contraction with filling (inflating) and relaxation with blowing—when the heart drives out the stream of blood.[26] By

---

[22] Willis 1664/81, p. 131.

[23] That the heart, as a muscle, pumps in series is an inference from Willis, ibid., p. 163, who stated that 'the blood … [is] emptied out of its right bosom into the Lungs, and consequently out of the left [bosom] into the Aorta'. For the tradition of auricles as bellows, see L. G. Wilson, 'Erasistratus, Galen, and the pneuma', *Bulletin of the History of Medicine* (1959), 33: 293–314.

[24] 'For 'inkindling' ('infiring'), see Willis 1664/81, pp. 95, 126, 129, 131, 136, 152, 154, 162 and unpaginated table, 'Accension, *An Inkindling*'; see also 'boiling up' ('ebullition', 'deflagration') of the blood, ibid., pp. 84, 90, 96, 104, 107, 124, 131, 155 and unpaginated table, 'Ebullition, *A boiling up*'.

[25] Ibid., p. 135 and unpaginated table, 'Systole, *Is the motion of the heart and arteries, contrary to Diastole, by which they are contracted, but by this they are dilated*'. For the belief that muscle increases in volume on contraction, see E. Bastholm, *A History of Muscle Physiology*…(Copenhaven, 1950); see also Kassler, *Inner Music* (note 7), pp. 188–192.

[26] Willis 1664/81, pp. 119, 124, 126, 129, 131, 135-136, 152–155, 162, 165–166 and unpaginated table, 'Inflated, *Blown or puffed up as a Bladder with* wind'; see also ibid., pp. 129, 135–136,. 152, 154, 175 and unpaginated table, 'Explosion: *A driving forth with violence*', 'Exploded, *Thrust forth, or driven out*'. In Willis's usage, blowing up denotes inflation and explosion signifies blast, as when a forcible stream is driven out by a bellows. Many commentators on Willis have misunderstood this usage; see, e.g., R. G. Frank Jr., *Harvey and the Oxford Physiologists: Scientific Ideas and Social Interaction* (Berkeley, 1980), pp. 221–223 *et passim*; R. Hierons, 'Willis's place in the history of muscle physiology', *Proceedings of the Royal Society of Medicine* (1964), 57:

these motions the two halves of the heart maintain the blood's flow at constant pressure as well as at constant temperature.

In water organs the hydraulic mechanism operates in conjunction with air feeders, but in some organs these feeders are air pumps and in others, wind bellows. In Willis's human *hydraulis* the equivalent of air feeders are the lungs. These two respiratory organs, placed on either side of the heart, communicate with the central windpipe that includes the bronchia, the trachea or windpipe proper, and the larynx or voice organ with its associated strands of cartilage and elastic ligaments. But the 'act' of respiration is determined by the hydraulic mechanism, since 'the Lungs themselves ... which the blood, boiling out of the Heart, passes through', are 'altered every minute of an hour almost' by the temper of the blood and 'the tenour of the Pulse'.[27] Hence, like air feeders in hydraulic organs, the lungs operate in conjunction with the hydraulic mechanism.

When Willis was writing *Cerebri anatome*, knowledge about respiration was in process of transformation, and the new discoveries do not appear in that work.[28] Instead, Willis attempted to explain why it is that respiration, though involuntary, may sometimes be 'restrained, interrupted, and divers[e]ly altered' at our pleasure.[29] He

687–692; and J. A. Bennett, 'A note on theories of respiration and muscular action in England c.1660', *Medical History* (1976), 20: 59–69, p. 65.

[27] Respiration: Willis 1664/81, p. 155; see also pp. 111–112, 114–115, 123, 136, 150–155, 163–165, 174–176.

[28] In 1661, with the aid of a microscope, Marcello Malpighi established the existence of an alveolar-capillary system, thereby scientifically completing Harvey's doctrine of the circulation of the blood. Then, in 1664, Robert Hooke blew air through the trachea of a dog with the thorax displayed, thereby proving that a constant source of fresh air was essential for life, that the mechanical movements of the lungs were of secondary importance and that inflating the lungs of a dead or dying animal will put the heart in motion. By this experiment, too, the ground was laid for understanding how air got into the lungs, what use it was to the blood when it got there, and so on. See L. S. Gottlieb, *A History of Respiration* (Springfield, Illinois, 1964), pp. 6–26, 45–46 and W. P. D. Wightman, *The Emergence of Scientific Medicine* (Edinburgh, 1971), pp. 74–95.

[29] Willis 1664/81, p. 155. In the medical tradition that Willis inherited, a distinction was made between three classes of movements: vital, natural and animal. Vital were considered involuntary and animal, voluntary, whereas natural were those movements partly under our control and partly not.

rejected the common and long-held opinion that the lungs are 'merely passive' instruments on the grounds that there are many branches of the nerves bestowed on the lungs, so that the lungs feel 'the first instincts of their motions, and ... do in some measure exercise themselves, and endeavour the *Systole* and *Diastole*, and design them according to the sense of its proper necessity'.[30] But when the nerve fibres requisite for local motion are wanting, the motive endeavours of the lungs are assisted by the diaphragm and muscles of the thorax; and by the 'sympraxis' or consent of all these parts, 'compleat breathing is effected'.[31]

Since Willis specifically noted that the motions of the lungs 'imitate the motion of the heart' and since he defined respiration as 'a breathing forth or venting', the lungs must be wind bellows, illustrated in *Figure 6.5*, whereby air is inspired and expelled by the lungs.[32] But the bellows structure includes more than the lungs, as Willis made clear, when he noted that the ring-like cartilages of the larynx and trachea have a 'double' or bellows-like motion not only for inspiration and expiration but also for modulating the voice, making it flatter or sharper, louder or softer. Hence, although the larynx and trachea are part of the central windpipe, they are part of the bellows structure and not an organ pipe, since, according to Willis, the cartilages of both are drawn 'hither and thither, like the folds of a pair of Bellows, both for breathing and [for] making a sound'.[33]

---

[30] Willis 1664/81, p. 155, who also rejected the received opinion that 'this reciprocation of the Lungs doth proceed wholly from the motion of the *Thorax*, and doth obey or observe its dilatation or constriction with a certain necessary dependency; viz. that the Breast being dilated or spread open after the manner of a pair of Bellows doth compel the ambient Air into the *Trachea*, which rushing into the Lungs, blows up and distends them; then the same Breast subsiding or sinking of it self, that the Lungs being pressed together with the weight of it, do breath[e] forth the Air before intruded'.

[31] Ibid., unpaginated table, 'Sympraxis, *A joynt exercising or agreement in practice or action; a consent of operation*'.

[32] Ibid. (lungs as bellows), pp. 153, 155.

[33] Ibid., pp. 150, 154.

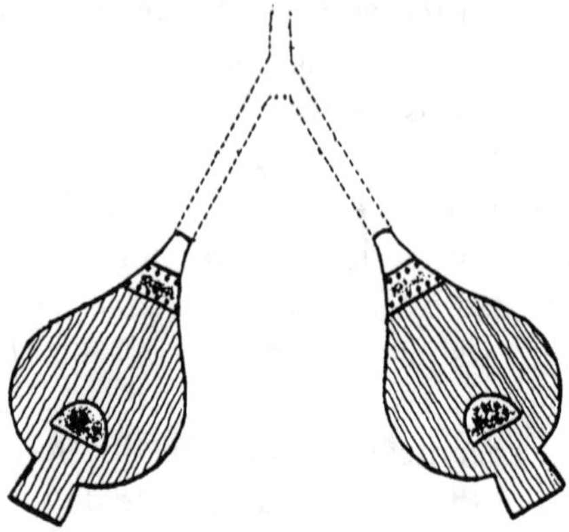

Figure 6.5. Pair of wind bellows

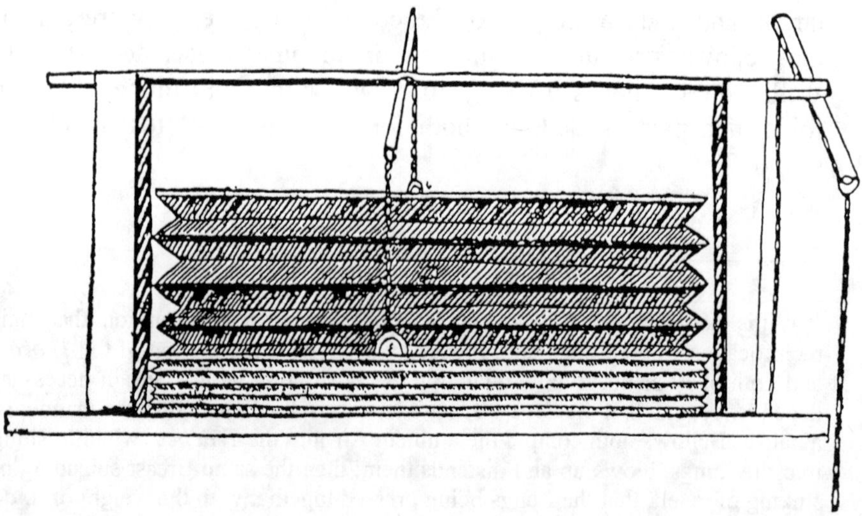

Figure 6.6. Lantern bellows

To function as wind bellows, the lungs require a pumping apparatus, so that they may be inflated to their full extent. In Willis's human *hydraulis* the pump is the thorax (see *Figure 6.3*), a framework of bones and soft tissues bounded by the diaphragm below, the ribs and sternum in front, the ribs and thoracic portion of the vertebral column behind, and above by the structures in the lower part of the neck.[34] Within the thorax are situated the precordia or parts about the heart, including the pericardium and the pleura. The former, which is attached to the diaphragm, is the serous membrane that envelops the heart, and the latter, another serous membrane, invests the lungs and lines the internal thoracic cavity.

When the thorax is increased in size, air goes into the lungs in inspiration; and when the thorax retracts, air is forced out in expiration. But sometimes the lungs, as well as the 'exterior Organs' of respiration, may influence the motion of the thorax. When this happens, there is a consent between the parts of the blowing mechanism, whereby 'all the Organs of Respiration intimately conspire and agree among themselves, that although one of them do a thing inordinately, rather than there shall be a Schism, the rest do imitate or follow its irregularity'.[35] These statements suggest that the thorax itself is like a lantern bellows, illustrated in *Figure 6.6*, a type that moves every way and yet continues parallel, thus allowing for irregular movements, as when the diaphragm 'begins laughter, the Lungs perform the same with a following cackling sound'.[36]

---

[34] Ibid. (thorax as bellows pump), p. 163, where Willis remarks that 'the *Diaphragma*, instigated by the passage of the nerves going out of this infolding, is drawn upward by a more rapid *Systole*, and raises up the Lungs, as it were making iterated leaps, and causes them, by their more frequent striking together, to drive out both the Air and the blood'.

[35] Ibid., p. 155. There is a sympraxis of parts in both regular and irregular movements, so that, ibid., p. 175, 'the *Diaphragma*, with the Muscles of the *Thorax* and the parts of either, conspire in their motion with the action of the Lungs and Heart: and that between all these … a joynt action may be sustained'.

[36] Ibid., p. 155. Because laughter hinders respiration, it was of particular interest to physicians generally, including Willis's assistant, Richard Lower, who wrote:

> In laughing the diaphragm seems to be driven up into the thoracic cavity by the abdominal muscles forcing the viscera against it, and to give way to this force in jerks, as if with the object of driving air through the parts

In both water and pneumatic organs, bellows may supply wind either directly to a regulator closet under a row of pipes or indirectly via a cistern. That wind for the human *hydraulis* is supplied by the latter method becomes clear from Willis's assumption that circulation is a distillatory process: the blood, during its circuit, 'waters' the cortical substance of the encephalon, from which a volatile liquor, extracted from the blood, is sent into the medullary substance for further purification. During this part of the process, the liquor becomes spirituous, its viscosity preventing the spirits from evaporating or 'flying away'.[37] The encephalon, therefore, is one of the cisterns in the body. But, as we shall see, it is a double cistern, in which the bloody liquor becomes even more spiritous; and it is this liquor, with its animal spirits, that is sent from the encephalon into the nervous system by the force of the heart and pulsation of the arteries.[38]

As the control system, the encephalon incorporates three principal mechanisms (see *Figure 6.2*): the 'medullar Trunk', the cerebellum ('Cerebel') and the cerebrum ('Brain proper'). Although

---

> and so producing a laugh in the larynx. Hence, as the diaphragm is attached to the pericardium, it pushes and forces the heart and its base close to its own vessels, both in the region of blood-inflow and in that of blood out-flow. The doors of the heart being closed, there is, therefore, a temporary intermission of the circulation. This is clear from the generalized venous swelling which occurs in the neck, face, and forehead. As soon, however, as the laugh finishes, and the return of the diaphragm to its proper level pulls the heart down, so that it can recommence its systole and diastole, the circulation is restored, and the venous swelling, which was previously visible while the laugh was in progress, disappears completely with the re-emptying of the vessels.

See R. Lower, 'A Treatise on the Heart [1669]...' tr. by K. J. Franklin in R. T. Gunther, *Early Science in* Oxford (15 vols., Oxford, 1923–67), vol. 9, pp. 90–91; see also p. 145, where Lower observed that not only heart function but also cephalic function may be affected by laughing, because 'the blood cannot … in laughing … be poured or driven back from the sinuses [of the brain] into the cerebral veins'.

[37] Willis 1664/81, pp. 87–90 *et passim*. In addition to the greater circulation, there also are lesser circulations in various cisterns, cephalic and non-cephalic.

[38] Ibid. (double cistern), pp. 89, 91, 93, 95, 101, 108, 111, 113–116, 119–120, 124–126, 132, 149 *et passim*. Willis, ibid., p. 90, rejected the ancient notion, still held in his day, that the meninges ('meninxes', 'maters') contract and, hence, pump out animal spirits.

the cerebellum is attached to the hinder part of the cerebrum, these two mechanisms have no immediate communication, one with another; but they do have mediate communication by means of the medullar Trunk. This is not the same as our brainstem—that agglomeration of protuberances and swellings above the spinal cord and below the cerebellum, for in Willis's anatomy the Trunk not only includes the oblong marrow but also the corpora quadrigemina (four 'orbicular prominences'), the pons (one 'annular protuberance') and the complete spinal marrow.[39]

As the oblong marrow, the Trunk originates in the Brain where the corpus callosum ends and the corpus striatum begins, as illustrated in *Figure 6.7*. The latter body forms the top of the Trunk,

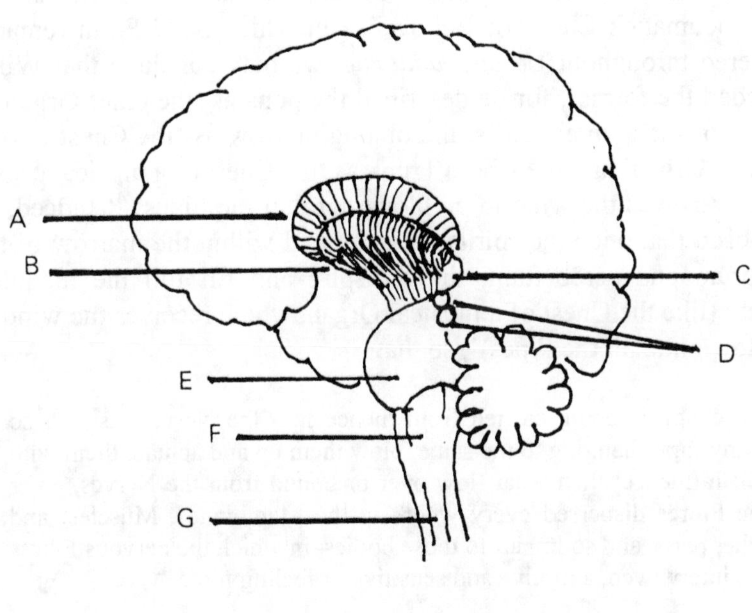

Figure 6.7. Pathway of medullar trunk
(A) corpus callosum, (B) corpus striatum, (C) thalamus,
(D) corpora quadrigemina, (E) pons,
(F) oblong marrow, (G) spinal marrow (cut off)

---

[39] Ibid., pp. 101, 121, 123.

which then extends past the thalami ('optic Chambers') into the corpora quadrigemina and pons. From these swellings and protuberances it enters the spinal cord and extends to the os sacrum, the ridge on which the backbone rests (for the os sacrum, see *Figure 6.1*). The medullar Trunk, therefore, is a common trunk to which the cerebrum and cerebellum are attached, so that, to Willis, it seemed 'like the Pneumatick Chest, or Bellows of a pair of Organs, which includes the blast or breath destinated to every Pipe; for in like manner the animal Spirits are contained in this marrowy tract, which blow up and actuate all the Nerves hanging thereto as occasion serves'.[40]

The passage just quoted contains two problems, one of which is the meaning of the term, 'pair of organs', to which I shall return. The other problem arises from Willis's statement that the Trunk is like 'the Pneumatick Chest, or Bellows'. But which is it? From remarks scattered throughout *Cerebri anatome*, we may conclude that Willis intended the former, for he described the pons as 'the chief Organ or Chest' of the animal spirits, the oblong marrow as 'the Chest ... of a musical Organ' and the whole Trunk as 'the Chest of a musical organ, which receives the wind to be blown into all the Pipes'.[41] Indeed, he supposed that once the spirits are sublimed within the marrow of the cerebrum and cerebellum, they 'inspire and fill full the medullar Trunk: (like the Chest of a musical Organ, which receives the wind to be blown into all the Pipes)', so that

> those Spirits being carried from thence into the Nerves, as into so many Pipes hanging to the same, blow them up and actuate them with full influence; then what flow over or bound from the Nerves, enter the Fibres dispersed every where in the Membranes, Muscles, and other parts, and so impart to those bodies, in which the nervous Fibres are interwoven, a motive and sensitive or feeling force.[42]

Since, in organs, the pipe rack is attached to the windchest, we should expect the same arrangement in the human control system.

---

[40] Ibid., p. 124.

[41] Ibid. (windchest), pp. 122, 126, 130.

[42] Ibid. (pipes), p. 126.

And we are not disappointed, for Willis pointed out that, as to 'the Nerves, ... all of them are produced immediately out of the medullar Trunk or its processes'.[43] Although sometimes Willis described the nerves as cords, strings or threads, and although he stated that spectacles or a microscope could not discover any cavity in the nerves, he himself continued to hold the ancient belief that bodily instruments contain *poroi*, that is, hollow channels. Hence, he described the nerves themselves, including their smaller parts, as 'Tube-like', thereby allowing the spiritous liquor to flow like wind in organ pipes.[44]

Once the pressurised air has reached the windchest of an organ, its access to the pipes is controlled by the key-frame. This mechanism is built up on the pallet chest, a long, narrow and shallow box, combining a soundboard, pipe rack with pipes, pallets and the key-frame itself. In organs having more than one rank of pipes, the soundboard is separated from the pallet chest by an intermediate board drilled with wind-ways for extra rows of pipes. When a key is depressed, a pallet running between the top of the windchest and the foot of the pipe moves into position, and the air under pressure passes through to the pipe foot. When the key is released, a strong spring pulls the pallet back to its first position, and the wind is cut off. The pallet, therefore, acts as a hinged slider or stop to control the admission of air from the windchest into the pipes.

If the nerve pipes are like organ pipes, they must have valves, the equivalent of pallets. That such valves exist is suggested by Willis's description of his eighth cranial nerve—the vagus, which originates as a bundle, each nerve in the bundle extending to different and distant parts of the body.[45] To maintain each nerve in the bundle

---

[43] Ibid. (pipe rack), p. 127.

[44] Ibid., p. 127; see also pp. 103, 132. For historical background, see G. E. R. Lloyd (ed.), *Hippocratic Writings* (Harmondsworth, 1978), pp. 84, 144–145 and E. Clarke, 'The doctrine of the hollow nerve in the seventeenth and eighteenth centuries', *Medicine, Science and Culture* ed. by L. G. Stevenson and R. P. Multhauf (Baltimore, 1968), pp. 123–141.

[45] According to R. Y. Meier, '"Sympathy" in the neurophysiology of Thomas Willis', *Clio Medica* (1982), 17: 95–111, Willis traced out the anatomy of the vagus ('wandering') nerve in much greater detail than his predecessors.

as a 'distinct channel', therefore, Willis supposed there were 'little doors in the middle of the branchings out of the nerves, as are in the Pipes of a musical Organ'.[46] Although this seems to be his sole reference to such 'little doors', we may assume that the vagus is not unique and that other nerve bundles also have hinged valves that function like pallets.[47]

While pallets are part of the control mechanism, they are themselves subject to control in one of two ways: either manually, by the action of a musician depressing and releasing one or more keys on the key-frame; or automatically, by clocking a barrel. In the latter instance, the action of pins on a revolving barrel can be transmitted by simple levers over some distance, so that if these levers come into contact with the pallets of organ pipes, the pipes will sound for as long as the pins remain touching the levers, and the length of the pipe tones will vary according to the width of the pins. To sound different pipes in the register, the pins are placed slanting on the barrel at regular intervals, so that more than one pipe can sound at once if there are several pins in a line along the axis of the barrel.

Since Willis held that 'the fourth, fifth, sixth, seventh, and eighth pair' of cranial nerves are employed 'for the performing the tacit Edicts' of the cerebellum for every involuntary function and that the remaining pairs perform the edicts of the cerebrum for voluntary function,[48] does this dual control mean that Willis's 'pair of organs' denotes a double organ, that is, an organ with two keyboards? To answer this question and, hence, to solve the other problem mentioned before, let us begin with the cortical substance of the encephalon, the pores and passages of which are filled with blood vessels.[49] These vessels are hidden by the 'barky' covering of the

---

[46] Willis 1664/81, p. 149 ('little doors'), and perhaps p. 159 ('stops'); see also unpaginated table, 'Valves, *A part of the brain, made like folding doors so called*'.

[47] For other nerve bundles ('bands', 'infoldings', 'trunks'), see ibid., pp. 126–127, 141, 150–151, 157, 162–163, 175, 179.

[48] Ibid., p. 121. Willis's classification of the nerves is not the same as our modern one.

[49] Ibid., pp. 88, 112.

cerebrum, which has many convolutions.[50] Because the convolutions are arranged asymmetrically, like the keys on the key-frame illustrated in *Figure 6.8*, Willis supposed that the cerebrum was suited to performance by 'the chief mover in the animal Machine', whose actions are free and changeable.[51]

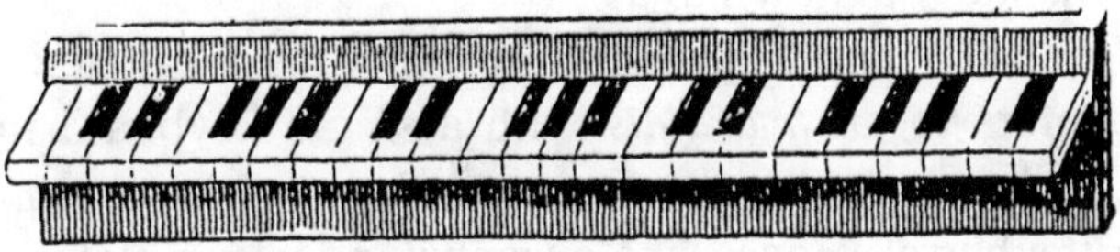

Figure 6.8. Key-frame

The cortical substance of the cerebellum differs from that of the cerebrum, for the blood vessels are hidden by 'lappets or little circles',[52] disposed symmetrically 'in a certain distinct series' and 'marked with the same form and proportion', like pins on a clock organ's barrel illustrated in *Figure 6.9.*[53] Because such regularly-

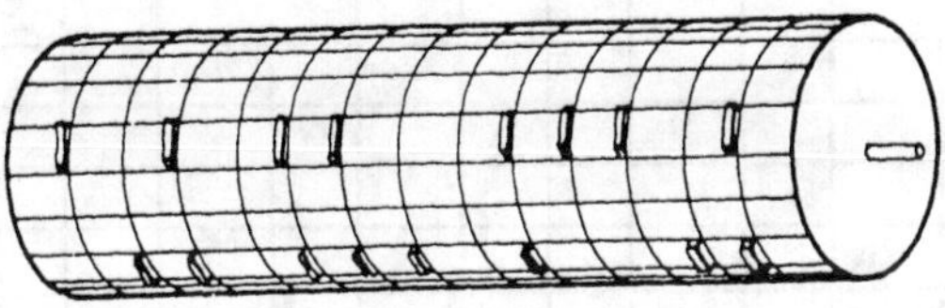

Figure 6.9. Pinned barrel

---

[50] Ibid., pp. 67, 92, 103, 111, 124. Although Willis, ibid., pp. 91, 124, described the two hemispheres of the cerebrum, he explained their duality on the grounds that if one hemisphere is defective, the other 'supplements', i.e., takes in twice the amount of bloody liquor from the cortex, the grey matter of which 'unfolds' into the cerebral cistern and its windchest.

[51] Ibid., p. 91.

[52] Ibid., pp. 67, 110, 112, 123.

[53] Ibid. (cerebellum as clock organ), pp. 111, 119.

disposed lappets are found in animals as well as humans,[54] the cerebellum controlled ‘certain necessary offices’ performed ‘in all after the same manner’ and which could be no other than ‘the motions and actions of the *Viscera* and *Præcordia*’, including ‘the beating of the Heart, easie Respiration, the Concoction of the Aliment, the protrusion of the Chyle’ and many other such motions, ‘made after a constant manner unknown to us’.[55]

We may conclude, therefore, that Willis’s term, ‘pair of organs’, refers to an *hydraulis* with a single keyboard that may be activated either manually, by fingering the keys, or automatically, by clocking the barrel. Hence, the encephalon, or complete control unit, may be thought of as a matrix, illustrated in *Figure 6.10*, in which the depression of a key, or the projection of a pin on the barrel, admits air to a specific column and the drawing of a pallet admits air to a

Pallets													*Nerves*
													etc.
													10
													9
													8
→	×				×								7
													6
													5
													4
→	×				×								3
													2
→	×				×								1
Keys/ Pins	C	C#	D	D#	E	F	F#	G	G#	A	A#	B	*Cortical Cells*

Figure 6.10. Encephalon as organ matrix

[54] Ibid., p. 87.

[55] Ibid. (cerebellum’s pre-determined programme), p. 111; see also pp. 109, 114–115, 117, 121, 123, 143.

specific row, so that any pipe, lying at the intersections of an open row and column will sound. Regardless of whether the control unit is on automatic or manual, change of pitch is achieved by using a separate generator, either a pin on the barrel or a key on the keyboard. If we limit the rank of pipes to Willis's ten pairs of cranial nerves (remembering, however, that the spinal nerves are also attached to the windchest), we may represent the encephalon's separate generators as different pitches in an octave, and the hinged valves as ten pallets or stops. We may then show by arrows and crosses that the pitches, C and E, are playing or being played, with the pallets drawn for the fourth, eighth and tenth pair of nerve pipes.

ORGAN FUNCTION

Although the mechanisms constituting the human *hydraulis* are so constructed as to provide a constant and perpetual wind, the source of this motive power is water pressure, namely, the blood that circulates round the body. During this process an oily humour, enkindled in the blood within the heart and lungs, is circulated with the arterial blood; and a spiritous liquor, distilled from the blood in the cerebral and cerebellar cortex, is circulated and further sublimed within the grey matter of the encephalon.[56] In treating these two products of distillation, Willis drew on the two-semen theory of Hippocratic doctrine, according to which female and male semen exist in all persons, regardless of sex.[57] In Willis's proto-chemical version of this

[56] Two sources for Willis's 'chemistry' seem important, though largely overlooked. One is Tertullian, who used the water organ as a symbol of Christian baptism; see C. A. Wilson, *Philosophers, Iōsis and Water of Life* (Leeds, 1984), pp. 58–59. The other is Bacon, who relied on 'pneumatic' explanations; see G. Rees, *Francis Bacon's Natural Philosophy: A New Source*...(Chalfont St. Giles, 1984), p. 39; see also J. C. Gregory, 'Chemistry and Alchemy in the Natural Philosophy of Sir Francis Bacon, 1561–1626', *Ambix* (1938), 2: 93–111; and M. West, 'Notes on the Importance of Alchemy to Modern Science in the Writings of Francis Bacon and Robert Boyle', *Ambix* (1961), 9: 102–114.

[57] For ancient doctrine, see Lloyd (ed.), *Hippocratic Writings* (note 44), pp. 317–346.

theory, the oily humour, or female semen, is the matter, and the spiritous liquor, or male semen, is the form of nourishment. Being in sympathy, the two types of semen commingle or 'marry together', thus imparting the 'energy' or 'plastic virtue' that blows the human *hydraulis*.[58]

Willis's doctrine of circulation, therefore, is not the same as that of William Harvey.[59] Indeed, the two men held different views about the heart, the valves and the blood. Harvey argued that the pulsific force of the heart is intrinsic to muscle fibre, for it is a relation between stretch and tension (contractility). He observed that, although the heart maintains pressure in the vessels, it is the valves that mechanically direct the flow of the blood. And he insisted that living blood, in circulating, brings its vital principle to the parts where it exercises their special functions. By 'vital principle' Harvey denoted heat, spirit and blood as one unit, for he assumed that blood became particulate only in the dead, when, as cruor, it separated into fibrous, gelatinous and serous fractions.[60] Though constantly changing in vitality and function, living blood did not change its physical properties—it did not become vaporous. Harvey, therefore, condemned spirits as a *deus ex machina* and the recourse to them as an asylum of the ignorant.[61]

---

[58] Willis 1664/81, pp. 89, 129, 135 ('our Hypothesis', 'a Paradox and very abstruse'); see also unpaginated table, 'Energy, *The force, or operation, or virtue of a thing*', and 'Plastic, *Formative, or that worketh and formeth*'. For sympathy, see below (note 71). Like the alchemists' *archeus*, Willis's blood is capable of protean transformation; and like the alchemists' purifications, Willis's transformational processes are distillations: circulation, separation, extraction, reduction, sublimation, conjunction and generation, during which the blood is dissolved in order to regenerate it in purified form as spirits (*aqua vitae*).

[59] W. Harvey, *The Anatomical Exercises: De motu cordis and De circulatione sanguinis* tr. by R. Lowndes [1653] and ed. by G. Keynes (New York, 1995).

[60] Ibid., especially p. 157.

[61] According to Harvey, ibid., p. 155, 'it is no wonder if Spirits, whose nature is left so doubtfull, do serve for a common escape to ignorance: For commonly ignorant persons when they cannot give a reason for any thing, they say presently, that it is done by Spirits, and bring in Spirits as performers in all cases'. But Willis replied, 1670/83, p. 23: 'I shall say nothing to those, who

According to Willis's doctrine, the fibres of the heart, or any other muscle, have no intrinsic property of contractility, because contraction is due to a chemical process that causes the blood to boil, the vapours to ascend and the heart to inflate.[62] There seem to be few, if any, strategically placed valves, because the peristaltic motion of the innervated vessels maintains circulation.[63] And living blood, as the liquid to be distilled, changes its physical properties: the distillate is the portion that evaporates when heat is applied and that is subsequently condensed by cooling; and the distillation residue is the portion that does not evaporate.[64] Hence, just as distillation separates evaporable substances from non-evaporable or less easily evaporable ones, so too does circulation function in the same way, for it is a process that separates substances according to their volatility.

In observing that there was a greater richness of blood supply to the grey nervous matter than to the white, Willis concluded that volatile spirits were distilled in the grey matter of the cerebrum and cerebellum and that the white 'medullar' matter of the Trunk or windchest formed merely their paths of travel. From the windchest originated the cerebral, cerebellar and spinal nerve pipes, the seats of reflex action.[65] Willis gave numerous instances of such actions, with special attention to the automatic acts of singing, laughing and weeping.[66]

---

wholly deny these Spirits, for ... the exi[s]tencie of which, is almost palpable, and may be proved demonstratively by the effects.'

[62] For systole, see above (note 25).

[63] Willis 1664/81, especially unpaginated table, 'Peristaltick, *Motion, a certain motion compassing about, as in certain Convulsions*', i.e., in peristalsis the vessels are 'pulled and hauled'. This motion affects not only all blood vessels (veins, arteries and 'bosoms', i.e., auricles and ventricles of the heart and the 'Choroeidal Infolding') but also, ibid., p. 179, all other vessels, lymphatic and nervous. In 1808 Thomas Young overturned the notion that the peristaltic contraction of the walls of the arteries was an important cause of circulation; until then, however, the notion was widely held.

[64] For cephalic distillation, see above (notes 37 and 58).

[65] I.e., 'act of reflection'.

[66] Ibid., pp. 117, 119, 141–143, 154–155, 163, 175; see also Willis 1670/83, pp. 80–81.

In automatic singing, for example, a stimulus is promptly followed by movement without conscious participation of the will. The spirits run up the sensory nerve pipe, reach the windchest and thence run down the motor nerve pipe. The higher up towards and into the grey matter of the cerebrum a reflex action occurs, the more does conscious mind attach to it. According to ancient as well as Cartesian doctrine, the cerebellum was the seat of memory.[67] Willis departed from this doctrine by locating the seats of innate and acquired memory in the cerebellar and cerebral cortex, respectively.[68] In this way he could conceive the cerebellum as a cistern for spirits that control reflex actions as well as reinforce the motor powers of the cerebrum.

For organ harmony there must be a strict sympraxis between the mechanisms of the operating system; and this requires regulation by spirits that blow in a smooth, even fashion.[69] These gentle winds are the vital instincts (innate impulse) that produce regular, unforced motions, for example, of the diaphragm in respiration. But intrinsic, as well as extrinsic, causes may excite spirits into a short 'fury' or a longer 'rage', so that they blow either as tempests or as whirlwinds. These vehement, uneven winds are the natural and preternatural passions (acquired impulse) that interrupt organ harmony either temporarily or for a longer duration, thereby producing forced and irregular motions, for example, of the diaphragm in laughing.

Willis believed that irregular motions are 'much more signal' in humans than in animals, because in the former alone there is a

---

[67] R. Descartes, *Treatise of Man* [1662]...tr. by T. S. Hall (Cambridge, Mass., 1972), who localised common sense and imagination in the cerebral ventricles and reason in the pineal gland. Against Descartes, Willis 1664/83, pp. 97 and 106, argued that the ventricles are 'sinks', and the pineal gland, a reservoir that receives and retains 'serous humors deposited from the arterious blood, till the Veins being emptied … sup them back'.

[68] Dual memory: Willis 1664/81, p. 188; for Willis's double cistern doctrine, see above (note 38). Until the work of Franz Joseph Gall early in the nineteenth century, most physicians regarded the cortex as merely vascular tissue. Willis was one of the exceptions to this rule.

[69] Ibid. (organ harmony), especially unpaginated table, 'Symphony, *A consent or agreement in harmony*', hence, different from sympraxis, for which see above (note 31).

'wonderful' sympraxis between the precordia and the face.[70] He attributed this consent to the conformation of the 'intercostal' or great sympathetic nerve, which reaches into the precordia and the viscera of the whole abdomen and continues by its superior ramification into the eyes and parts of the mouth and face.[71] Hence, in every passion the eyes, face and mouth correspond with the affections of the precordia 'often unknown to us or against our minds', so that we automatically betray 'the most intimate sense of the heart by the countenance and aspect'.[72] Indeed, because many shoots of the intercostal nerve are sent to the nerve of the diaphragm, Willis claimed that 'risibility is the proper affection of man'.[73]

In summary, we may say that Willis's human *hydraulis* effects vascular work as water work and nervous work as wind work. But the source of this work is the blood, an hydro-pneumatic principle, conceived as 'chemical work',[74] the distillatory processes of which occur in various reservoirs and cisterns, including the encephalon itself, which sublimes spirits from the bloody mass, 'inspires' the windchest and controls the entrance of spirits into the nerve pipes, as well as into the nerve fibres that innervate the muscles. In short, nervous work is performed by the products of distillation—the vapours or breaths that Willis called 'spirits'.

---

[70] Willis 1664/81, pp. 174–175.

[71] Ibid., unpaginated table, 'Sympathy, *Is a natural passion of one thing to another, or an agreement in qualities*'; 'Sympathic, *Agreeing in affections or passions*'. For feelings of sympathy and antipathy ('molestation'), see ibid., pp. 109, 114–116. These feelings, which arise from the agreement or disagreement in 'qualities' (i.e., the intrinsic elementary mixtures), are the basis of the harmony or disharmony that the involuntary nervous system brings about. This doctrine of mixture is overlooked in the otherwise excellent article by Meier, '"Sympathy"' (note 45), p. 103.

[72] Willis 1664/81, p. 117.

[73] Ibid., pp. 117, 163, 175 and unpaginated table, 'Risibility, *Laughter, or the faculty of laughing*', perhaps echoing Porphyry the Phoenician; see his *Isagoge* tr. by E. W. Warren (Toronto, 1975), p. 57 *et passim*. For a critique of Willis's supposition, see J. Mayow, *Medico-physical Works* [1674] (Edinburgh, 1957), pp. 201–202.

[74] Willis 1664/81, p. 88.

Although Willis admitted that spirits are invisible, he treated them as a heap of real, impenetrable little bodies, since he compared their constitution (*systasis*) to 'Atoms or subtil Particles ... chained and adhering mutually one to another'.[75] On this premiss he could argue two things. First, if spirits are substances, or that which subsists, they must be the essence (*hypostasis*) of the soul. And second, since that which changes is their qualities, attributes or accidents, we may imagine mutation (*metathesis*) as an interchange of little bodies or groups of little bodies, whereby spirits are 'figured together in a certain Species' or informing form.[76] Hence, just as wind is the life or internal energy of the hydraulic organ, so too are spirits the psychic life of the human *hydraulis*.

Psychic life, however, has non-rational as well as rational powers, each of which has distinct skills. To the rational 'soul' belong all 'pure Affections, and such as are Simple, coming without perturbation or trouble', whereas to the non-rational 'corporeal soul' belong 'all the vehement Affections or Perturbations of the Mind, by which it is wont to be moved, and inclined hither and thither, for the Prosecuting the Good, or shunning Evil'.[77] This kind of dualism may be traced to the Aristotelian tradition that would have been part of Willis's medical training.[78] Indeed, in *Cerebri anatome* there is considerable evidence that Willis continued aspects of that tradition. Like Aristotle, he believed that the affections can be controlled and modified by the rational power, since he argued that, even though sense impressions that set up a reflex may be perceived 'without our knowledge or intention', we may become aware of the resultant

---

[75] Ibid., p. 132.

[76] For *systasis, hypostasis* and *metathesis*, see ibid., p. 130; for species, see ibid., pp. 77, 92, 11, 117–120, 123, 127. Willis retained the Peripatetic doctrine of species ('forms', 'impressions', 'marks', 'schemes', 'types'), according to which movements like eddies represent the action of the transmuted 'power' of percussion; such action, though conceived as incorporeal, invisible and 'spiritual', by moving corporeal 'instruments' at any time becomes 'material'. For Hobbes's deconstruction of this doctrine, see Kassler, *Inner Music* (note 7).

[77] Willis 1670/83, pp. 42–43.

[78] See T. J. Tracy, *Physiological Theory and the Doctrine of the Mean in Plato and Aristotle* (Chicago, 1969).

muscular effect.[79] Moreover, he hinted that acts of prudence and virtue rely on such awareness, which is the ground for learning, by habituation, to control the reaction.[80] The affections, therefore, might be considered rational in that they are capable of obeying the rational power.

So far, then, Willis continued the Aristotelian tradition. But his adherence to this tradition is called into question when we consider the highest skill of the rational power. For Aristotle, that skill is speculative reason, the ability to know universals.[81] This knowledge relies on the non-rational power, since universals exist only in particular, concrete, sensible things. Aristotle's speculative reason, therefore, depends upon phantasms built up through successive sense impressions. For Willis, however, the rational power's highest skill is insight—self-perception—and this suggests Platonic beliefs.[82] In *Cerebri anatome* Willis had little to say about this skill. Instead, he focused on the skills of the non-rational, corporeal soul: vital, in which impulse is due to intrinsic (sympathetic) properties of the human *hydraulis*; natural, in which impulse is due to external stimuli; and animal, in which impulse is due to 'phantasie', a faculty directed to conscious and intentional action. Hence, we might ask: how is phantasie directed?

Willis answered this question in *De anima brutorum*, where he observed that the skill of phantasie is twofold. First, there is receptivity, since 'she' (note the pronoun) apprehends 'simple things' suggested either by instinct or by habit. Second, there is imagination, since she composes or divides 'many things at once', and from thence makes 'enunciations' or phantasms. But the psychic life of the body

---

[79] E.g., Willis 1664/81, p. 114: 'unless by the effect it [the feeling of sympathy] cannot be known'.

[80] Ibid., p. 162: 'Certainly the Works of Prudence and Vertue depend very much on the mutual commerce which happens to the Heart and the Brain [cerebrum]: because, that cogitations about the acts of the Appetite or Judgment may be rightly described, it is behoveful for the flood of the blood to be restrained in the Breast, and the inordinations of it and of the Heart itself to be governed by the Nerves … and to be composed into requisite and apt motions.'

[81] Aristotle, *Posterior Analytics* (100a59).

[82] Willis 1670/83, pp. 39, 41 *et passim*.

suffers and is corrupted by intrinsic as well as extrinsic causes. Hence, phantasie's skill must be moderated and controlled; and this is done by intellect, which 'beholds' all enunciations conceived by phantasie and, judging them to be true or false, orders and disposes them into a series of notions 'accommodated to speculation or practice'. When 'adorned by Learning', intellect 'sublimes' and divests phantasie of matter so as to form universals from particulars and to frame thoughts and speculations on such higher notions as infinity, eternity, angels, God and 'it self'. But intellect,

> even as it beholds all the Phantasms, and Orders and Rules them at its pleasure ... not only perceives, but whil'st it is its self, governs and moderates, all Concupiscences, and Floods of Passions, that are wont to be moved also within the Phantasie; and so, as it [intellect] approves these Affections, and rejects those, now excites others, no quiets them, or directs them to their right ends, the Rational Soul it self is said to exercise certain Acts of the Will or Power, by these kind of Dictates of hers [i.e., phantasie's enunciations or phantasms], and that she her self wills or wills not, the same thing, which by her Permission or Command, the Sensitive Appetite desires or hates.[83]

There are three points to note about Willis's doctrine of the soul. First, there is an identity between intellect and rational will, just as there is an identity between phantasie and sensitive appetite.[84] Willis, therefore, utilised these terms as different ways of understanding psychic processes. Phantasie is a shorthand term for the non-rational power or corporeal soul that mediates changes in both the operating and control systems of the human *hydraulis*. But there are two parts to this process, passive-sensory and active-

---

[83] Ibid., pp. 42–43.

[84] A doctrine of the identity of the will and intellect was commonly held by seventeenth-century Platonists in opposition to the received Thomistic doctrine of the real distinction of these powers. See D. W. Dockrill, 'Spiritual knowledge and the problem of enthusiasm in seventeenth century England', *The Concept of Spirit* ed. by D. W. Dockrill and R. G. Tanner (Auckland, 1985), pp. 147–171; see also G. A. J. Rogers, J. M. Vienne and V. C. Zarka (eds.), *The Cambridge Platonists in Philosophical Context* (Dordrecht, 1997).

motor.[85] During the first part of the process, the corporeal soul suffers either intrinsically derived stimuli—sympathies and antipathies—or extrinsically derived stimuli—pleasures and pains. But what follows, in the second part of the process, is a complex of affections, whereby the organ mechanisms are perturbed. The intellect is immediately aware of this perturbation and associates with it an opinion of good and evil. That opinion is an act of pure cognition and not, apart from the accessory suffering and perturbations, a passion or an affection. But in this synthesis, or 'conjugal compact', the psychic process is completed.[86]

Second, intellect and phantasie both occupy the same seat in the corpus callosum, the material substrate of imagination. Willis conceived this mental end organ as the dark chamber of a camera obscura with a double convex lens ('dioptric Looking-Glasses').[87] On this model, species or informing forms are 'intromitted', like rays of light, through sensory nerve pipes to the focus or corpus striatum ('objective Glass'), the material substrate of common sense. From that body they pass to the corpus callosum, where the species are represented as phantasms 'upon a white Wall'. If a phantasm imports only knowledge of an object, it is sent for storage from the corpus callosum to the cephalic cortex, the material substrate of memory. But if, from sense or from memory, the phantasm imports either good or evil, the species are reflected from the corpus callosum to the medullar Trunk or windchest. From that Trunk wind is admitted to one or more of the motor nerve pipes and nerve fibres implanted in the muscles, so that the good or evil may be embraced or removed.

---

[85] Willis 1664/81, p. 95: 'But there happen to this [corporeal soul], because it is apt to be moved with a various impulse, and so to contract or dilate its species [emanations] in whole, or in part, for that reason [there are] divers manners both of Actions and of Passions, to wit, the Senses, which we call its Passions; and Motions, which we name the Actions of the same.'

[86] Willis 1670/83, p. 47.

[87] Ibid., pp. 24–25, corresponding to the bodily eye, pp. 75–85. The conception of the eye as a camera obscura became commonplace after the demonstration by Johannes Kepler of the retinal image by means of a camera obscura with a lens. See, e.g., T. Browne, 'The garden of Cyrus' [1658], *The Major Works* ed. by C. A. Patrides (Harmondsworth, 1984), pp. 376–377.

Third, even though intellect beholds phantasms and employs rational will in making use of them, there is no identity between intellect and phantasie, because the one is created by, and has its gifts from, the deity, whereas the other is 'traduced' by, and has her gifts from, the male parent.[88] The latter gifts include, but are not limited to, the temperament or 'idiocrasie' of the male and female semen—the elementary particles that constitute the two chief humours in the blood.[89] Those seminal principles, having pre-existed from the beginning of time, 'rise up' to the predestined form of the body by the law of creation, not by a chance concourse of atoms.[90] Once inherited, however, they persist during life 'and have their times of crudity, maturity, and defection'.[91] Since the 'seed plots' produce innate predispositions, and 'custom' produces acquired dispositions, the two together account for differences between persons as well as in the same person over time.

Despite these differences, Willis identified three ways in which good and evil are brought to a person: either they respect the corporeal soul by herself, abstracted from any other relation; or they respect her as joined to the body that she 'cherishes'; or they respect

---

[88] Willis 1670/83, p. 42: 'Wit, Temperament, Ingenuity, the Affections, and other Animal Faculties … proceed immediately from the Corporeal Soul, which we grant to be begotten by the Father, together with the Body, but not the Rational Soul.'

[89] Willis 1664/81, unpaginated table, 'Idiocrasie, *The proper disposition or temperament of Bodies*'; see also Willis 1670/83, unpaginated preface. Although Willis's doctrine of mixture is proto-chemical, he couched it in terms of ancient, especially Galenic, doctrine, ibid., pp. 105–234. The elementary mixtures, whether inherited or acquired, affect both the contained fluids and the containers, which may be too wet, dry, cold or hot; the containers and the contained, in turn, affect the patient's 'complexions', which may be choleric, phlegmatic, melancholic or sanguinic.

[90] Ibid., p. 30, where Willis espoused a doctrine of special creation, when he wrote that the 'beginning of all things proceeds wholly from divine Providence, directing Generations, to the Ends and Ideas of Forms, according to the original Types primitively ordained by the same'.

[91] Ibid., pp. 204, 211, 216.

her as subdued by the rational soul.[92] But phantasie tends either to obtrude beyond 'the bounds of the Body' or to retire from the sphere of her 'Emanations', since in pleasure 'she stretches forth her self into a great Compass' whereas in pain she contracts 'more narrowly'; and from 'this twofold Affection ... all the other Passions [of the mind] take their Origine'.[93] Phantasie, therefore, is the unregenerate person, the inconstant, undisciplined human equivalent of intellect. Adhering to the flesh, she inclines to sensual pleasures and has to be helped by the intellect, either by means of ethical rules or by divine favour. Intellect, therefore, is Willis's supreme regulatory principle—the normative person conceived as a simple, unextended 'pure spirit', seated, during earthly life, in the mind's eye, like a musician seated in his organ room in *Figure 6.11*.

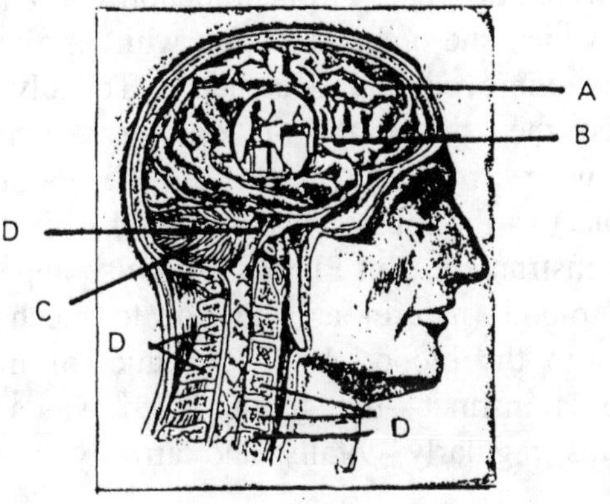

Figure 6.11. Encephalon, showing seats of the faculties
(A) key-frame/cerebral cortex ≈ acquired memory
(B) musician's room/corpus callosum ≈ phantasie and intellect
(C) barrel pins ≈ innate memory
(D) windchest with soundboard and pipe rack/
medullar trunk (cut off) ≈ reflex action

---

[92] Ibid., p. 46. From these three ways Willis arrived at a classification of the passions in general—physical, moral and metaphysical. Regarding the second class, Willis observed, ibid., pp. 42–43, that as the corporeal soul is 'nearer to the Body', she bears 'a more intimate Kindness or Affinity towards the Flesh' and is 'tied wholly to look to its Profit and Conservation'.

[93] Ibid., p. 48.

As Willis himself affirmed:

> one most subtilly Corporeal Soul, is joyned immediately to the Body, and is intimately united, and that by the intervention of this Soul, another immaterial, residing in its Bosom, inhabits the Body, and is the supream and principal form of the whole Man: But that after Death, the Corporeal Soul being extinct, this survives and is Immortal.[94]

Conclusion

From the foregoing we may conclude that Willis adopted a mystical approach to the passions. A person is both affected and not affected by a passion, which is a natural phenomenon that takes place in the person and of which she or he is aware, while at the same time the person remains unmoved and detached.[95] To solve this paradox, Willis accepted the trichotomy of Platonism, since he made a distinction between intellect (*mens*) as the immortal musician, phantasie (*anima*) as the musician's skill, and body (mechanism) as the musician's instrument. But Platonism is also apparent in Willis's conception of blood (nature) as superior to mechanism (matter), because life is in the blood. As the product of intrinsic seminal sympathy, life is instinct—the gentle, even winds that blow the human *hydraulis* regularly, vitally and artfully in order to attain

---

[94] Ibid., unpaginated preface. Willis's doctrine is close to that of Robert Boyle, who also conceived the living body as an 'hydraulico-pneumatical engine'. For Boyle, the intellect is capable of judging when illumined by 'the evidence of perception', i.e., insight or 'intuition', because judging is performed without 'instruments', i.e., without rules, hypotheses, theorems or conclusions in philosophy or divinity. Boyle also believed that intellect is so constituted that it has both 'inbred' and 'acquired ideas and primitive axioms … by relation or analogy whereunto it judges of all other notions and propositions'. See R. Boyle, *The Works* ed. by T. Birch (6 vols., London, 1965–66), vol. 5, pp. 225, 406, 717.

[95] H. M. Gardiner, R. C. Metcalf and J. G. Beebe-Center, *Feeling and Emotion: A History of Theories* (New York, 1937), p. 72, who identify five distinct approaches to freedom from subjection to the passions: Stoic, Peripatetic, common-sense, mystic and psychological.

ends.[96] But if intrinsic sympathy is perturbed, so too is the sympraxis between the organ mechanisms, as, for example, in laughing.

As noted previously, laughter may be instigated by the vehement, uneven winds of passion, natural or preternatural. In natural passion there is a fury and in preternatural passion, a rage of spirits in the corpus callosum. Unless restrained, therefore, these cerebral 'storms' excite phantasie's emanations to blow commensurately. In a fury there is a transient delirium or frenzy (delirious excitement), whereas in a rage there is a more vehement commotion that may be temporary (delirium proper) or longer lasting (frenzy proper).[97] Although differing in force and duration, all cerebral storms are disordered mental states of acute onset, characterised by less or more confusion, disorientation, disorders of perception (hallucinations and illusions), delusions, vigilance and over-activity of psychomotor autonomic nervous system functions. In the case of raging storms, intellect becomes alienated; and the physician is required to intervene.

Only when 'itself'—not alienated—can intellect, as rational free will, restrain phantasie, so that her emanations, no longer dancing 'like mad Bacchanals', perform regularly according to certain

---

[96] Henry More and others of the Cambridge Platonists propounded a similar doctrine. Relying on the authority of Scripture for their truth, these men, like Willis, sought to animate the Cartesian machine with Plato's *anima mundi*—what the Cambridge men variously named 'plastic nature', 'spirit of nature', 'hylarchic principle'. See, e.g., R. L. Colie, *Light and Enlightenment: A Study of the Cambridge Platonists and the Dutch Arminians* (Cambridge, 1957), pp. 55–57 *et passim*. Their connection to Willis seems to have been indirect, through Willis's patient and More's pupil, Anne Conway. Another patient, however, was Willis's direct source for scriptural authority. This was the Rev. Henry Hammond, whom Willis described as 'the glory of Learned Men'. For this description and for Hammond's medical case history, see Willis 1670/83, pp. 40–41, 224.

[97] For delirium and frenzy, see ibid., pp. 179–187. For the ancient tradition, see Lloyd (ed.), *Hippocratic Writings* (note 44), pp. 87–138, where the term 'delirium' signified a temporary or transient state of being out of one's mind. But the anonymous Hippocratic physician, ibid., pp. 103, 138, also distinguished between delirium without excitement, where the patient is silent, and delirium with excitement, the signs of which include, but are not restricted to, 'much talking, laughter, singing'.

'numbers, ways and measures'.[98] Like Ion, Plato's rhapsodist, phantasie's skill is artful by a divine allotment that is apprehended in a non-cognitive, non-sober and, therefore, 'mad' intuition (obscure imagination).[99] When so inspired, phantasie, like an automaton, becomes 'metaphysically' excited. Conveying her frenzy to the rest of her emanations, sympraxis is restored; and the organ mechanisms produce harmonious music, rather than cacophonous noise. Thus we may understand Willis's argument from design, that the music of the hydraulic organ 'deservedly amazes us, and we acknowledge this Effect, far to Excel both the matter of the Instrument, and of the hand of the Musician striking it'.[100]

APPENDIX

When considered in the context of the technology of his day, hydraulic organs, like Willis's model, commonly consisted of operating and control systems, each with its special 'motive complications'. Willis claimed to have seen water organs in action,[101] so that any of these could have been used as analogues for his *operating system.* But such organs did not commonly consist of a dual *control system*, one part of which works manually and the other, automatically.[102] We might ask, therefore, was Willis's theoretical model a figment of his imagination or was it based on an existing

---

[98] Willis 1664/81, p. 179; see also p. 119.

[99] Plato, *The Collected Dialogues* ed. by E. Hamilton and H. Cairns (Princeton, 1973), pp. 475–525; see also Chapter 1 *supra*.

[100] Willis 1670/83, p. 33.

[101] See above (note 15).

[102] W. Pagel, 'Medieval and renaissance contributions to knowledge of the brain and its functions', *The History and Philosophy of Knowledge of the Brain and its Functions*...ed. by F. N. L. Poynter (Oxford, 1958), pp. 95–114, p. 110, states that Willis relied on the hypothesis of William of Saliceto, a medieval surgeon, who held that the cerebrum governs voluntary and the cerebellum, involuntary motion, 'the reason being the composition of the latter of small uniform leaflets which because of their uniformity should produce automata in the form of robot-like animal spirits'.

analogue? Although no definitive answer can be given to this question, there was a unique organ that had features similar not only to parts of Willis's operating system but also and most notably to his dual control system.

Commissioned by Elizabeth I as a present for the Sultan of Turkey and built between 1597 and 1599, the result was what we now call 'an orchestrion', since it consisted of a combination of several instruments from different categories.[103] This organ was unique not only because it seems to have been the first orchestrion-type organ ever built but also because orchestrions were not built again until some 200 years later.[104] More importantly, the instrument included a clock with periodical automatic release of an organ that could be played either automatically or manually, at will, by touching a spring. Although the original specification is lost, a transcript was printed in *The Illustrated London News* for 20 October 1860, accompanied by a drawing, shown in *Figure 6.12.* According to this source, the operating mechanism consisted of three 'strong, forcible and artificial bellows, with a very strong, sufficient motion of wheels and pinions, and all other engines and necessary instruments and helps, very well wrought, and sufficient to drive and move the bellows at all times ... for the space of six hours'. And the control mechanism included a 'soundboard, with certain instruments or engines called his barrel and keys, and five whole stops of pipes'. By means of the three bellows, the barrel, the keys and 'other engines moving the said bellows', the pipes were 'to sound forth musically four or five songs without the hand of any person on the keys'. In addition to these mechanisms for the organ proper, others regulated the motions of a planetary clock, an hourglass, a chime of 'tuneable bells' and various figurines.

---

[103] Although the commissioned organ was pneumatic, not hydraulic, both types may be described as pneumatic, since only the system of blowing is different. In a water-blown organ, water is used to stabilise wind pressure, and this was the principle utilised by Willis; but in the air-blown organ, air is forced into a blast-bag by bellows or air feeders and then forced out of the bag either by pressure of the hand or by the action of weights.

[104] According to A. W. J. G. Ord-Hume, *Clockwork Music: An illustrated History of Mechanical Musical Instruments from the Musical Box...to the Orchestrion* (London, 1973), p. 194, the 'first' orchestrion was built by a clockmaker between 1789 and 1801.

Figure 6.12. The orchestrion

The whole was the result of a cooperative effort between the organ maker and Royalist, Thomas Dallam, and the horologist and goldsmith, Randolf Bull.[105] On its completion in 1600, Dallam took the organ to the Levant, and the diary recounting his travels and experiences in the Harem is still preserved.[106] In 1602 he returned to England, where he continued in his profession as organ builder until 1642, when he, with other members of his family, left England to set up business on the continent. Then, in 1660, after the Restoration, some members of the family returned to England to carry on the business. The returning members included two of Dallam's sons, Robert and Ralph, who settled in Oxford, where Willis was practising medicine until leaving for London late in 1667 or early in 1668. It is possible, therefore, that Willis could have heard of the Dallam-Bull organ from one of Dallam's sons.

Between 1661 and 1663 one son, Robert, planned and built an organ for New College, whereas some years later the other son, Ralph, was involved with the refurbishing of the 'publique Music Schoole' in the University, for which he provided '1 upright organ

---

[105] Thomas Dallam was a freeman of the Blacksmith's Company, whereas Randolph Bull, about whom little is known, probably was a member of the Goldsmith's Company, since the Clockmaker's Company was not formed until 1631. The contract for the organ was drawn up between Bull and Richard Stapers, governor of the Levant Company. This contract, along with the specifications for, and drawing of, the organ disappeared sometime after 20 October 1860, the date when the transcript appeared in *The Illustrated London News*.

[106] There is a brief description of the organ, as played for the Sultan ('Grand Sinyor'), as follows:

> Firste the clocke strouke 22; than The chime of 16 bels went of[f], and played a songe of 4 partes. That beinge done, tow personagis which stood upon tow corners of the seconde storie, houldinge tow silver trumpetes in their handes, did lifte them to theire heades, and sounded a tantara. Than the muzicke went of[f], and the orgon played a song of 5 partes twyse over. In the tope of the orgon, being 16 foute hie, did stande a holly bushe full of blacke birds and thrushis, which at the end of the musick did singe and shake theire wynges. Divers other motions thare was which the Grand Sinyor wondered at … [including] whye those keaes did move when the orgon wente and nothinge did tuche them.

See J. T. Bent (ed.), *Early Voyages and Travels in the Levant: 1. The Diary of Master Thomas Dallam, 1599–1600* (London, 1893), pp. 67–68.

with 4 stopps'.[107] This instrument, as well as the refurbishments, was paid for from funds collected during a period of ten years, beginning in 1665; and the benefactors included more than a dozen members of Christ Church, Willis's college, including the dean, John Fell (Willis's brother-in-law), Ralph Bathurst (Willis's medical colleague), Thomas Sprat and John Wallis. The three last persons, with Willis, were members of the Experimental Philosophical Club that met in High Street and sometimes in Wadham College and from which sprang the Royal Society.

[107] J. Hawkins, *A General History of the Science and Practice of Music* [1776] (2 vols., New York, 1963), vol. 2, p. 699.

7. THE SCIENCE OF MUSIC TO 1830

... the science of music, in its whole extent,
... is scarcely less intricate or more easily acquired than
the most profound of the more regular occupations of the schools... .
T. Young, 'An essay on music', *British Magazine* (1800)

INTRODUCTION

In the intellectual tradition stemming from the ancient Greeks and persisting to the seventeenth century, the term 'art' denoted skill and the term 'science' denoted the doctrine or theory of a skill. The arts comprised grammar, rhetoric, logic, arithmetic, geometry, astronomy and music. These seven arts, and their seven corresponding sciences, constituted the *enkyklios pædia* or traditional system of western knowledge and education.[1] The arts were referred to as mechanical if they were studied and practised for remuneration, whereas the very same arts were denominated liberal if they did not involve direct monetary advantage to the person studying and practising them.[2] In the early part of the sixth century A.D., Boethius coined the word

[1] See R. Collison, *Encyclopaedias...from 350 B.C. to the Present Day* (2d edn., New York, 1966).

[2] After the growth of learning in the twelfth and thirteenth centuries, the mechanical arts were differentiated in subject matter from the liberal arts. Possibly, the first to record this differentiation was Hugh of St. Victor, *The Didascalicon...A Medieval Guide to the* Arts tr. by J. Taylor (New York, 1961), who classed '*lanificium, armatura, navigatio, agricultura, venatio, medicina, theatrica*' as mechanical arts. For a survey of classifications of human knowledge from antiquity to the nineteenth century, see P. O. Kristeller, 'The modern system of the arts', *Renaissance Thought II: Papers on Humanism and the Arts* (New York, 1965), pp. 163–227.

'quadrivium' to designate the mathematical arts—arithmetic, geometry, astronomy and music;[3] and it is thought that this led by analogy to the term 'trivium' being used for the vocal arts or disciplines dealing with words instead of with numbers and magnitudes.[4] Hence, the encyclopaedia constituting the traditional system of knowledge may be represented as follows:

ENCYCLOPAEDIA OF ARTS/SCIENCES

Trivium (Vocal Arts/Sciences)			Quadrivium (Mathematical Arts/Sciences)			
Grammar	Rhetoric	Dialectic	Arithmetic	Geometry	Astronomy	Music

As an educational subject music is the oldest of the arts, for before the time of Plato and Aristotle, music and gymnastic were the staples of Greek education.[5] But the term '*musikē*' had a breadth of meaning now lost, for it signified *the* vocal art of declamation to which grammar, rhetoric, poetry and other literary studies were subservient musical arts. As the vocal art, music also embraced performance on the kithara and aulos, which were used to accompany sung recitations. With the cult of Pythagoras, music also came into prominence as a mathematical art, for the Pythagoreans believed that

---

3 For the variable position of music in the encyclopaedia, see E. A. Lippman, 'The place of music in the system of liberal arts', *Aspects of Medieval and Renaissance Music*...ed. by J. LaRue (New York, 1966), pp. 545–559.

4 L. Minio-Paluello, 'Anicius Manilius Severinus Boethius (A.D. c.480–524/525)', *Dictionary of Scientific Biography* ed. by C. C. Gillispie (16 vols., New York 1970–80), vol. 2, pp. 228–236. According to F. V. Hunt, *Origins in Acoustics: The Science of Sound from Antiquity to the Age of Newton* (New Haven, 1978), p. 11, the ancient Greeks regarded geometry 'as magnitudes at rest, astronomy as magnitudes in motion, arithmetic as numbers absolute, and music as numbers applied'.

5 See H. I. Marrou, *A History of Education in Antiquity* tr. by G. Lamb (London, 1956), p. 76; see also G. R. Morrow, *Plato's Cretan City: An Historical Interpretation of the Laws* (Princeton, 1960), p. 302, who states that although Greek music and dancing left no enduring monuments comparable to those of poets, sculptors and architects, 'the indirect evidence of the primacy of the musical arts is overwhelming and indisputable'.

number was the essence of all things, and they based this belief on three observations:

> The first of these was the mathematical relationships of musical harmonies—that is, that when the ratio of lengths of sound-producing instruments ... is extended to other instruments in which one-dimensional relations are involved, the same musical harmonies result. Secondly, the Pythagoreans noted that any triangle formed of three sticks in the ratio 3: 4: 5 is always a right triangle, whatever the length of its segments. Their third important observation derived from the fixed numerical relations of the movements of the heavenly bodies. It was thereby apparent to them that since the same musical harmonies and geometric shapes can be produced in different media and sizes by the same combination of numbers, the numbers themselves must express the harmonies and shapes and even the things having those harmonies and shapes. It could thus be said that these things—or, as they were later called, the essences ... of these things—actually were numbers. The groups of numbers that embodies the essence of a thing, and by which it might be reproduced were called ... 'words' ... a term that later came to mean 'ratio'.[6]

The ancient and medieval conception of the encyclopaedia was indebted to later Greek practice that was codified in Plato's theories of education. In those theories, enunciated in the *Laws* and *Republic,* Plato outlined a circle of studies appropriate to lower and higher education. Studies in lower education were designed for infants from

---

[6] K. von Fritz, 'Pythagoras of Samos (c.560–c.480 B.C.)', *Dictionary of Scientific Biography* ed. by C. C. Gillispie (16 vols., New York, 1970–80), vol. 11, pp. 219–225, who also states that early Pythagorean music theory was based on observations drawn from the kithara and aulos, the most widely used instruments in Greek society and that

> from these observations it was concluded that the most beautiful musical harmonies corresponded to the most beautiful (because simplest) ratios or combinations of numbers, namely the octave (2:1), the fifth (3:2), and the fourth (4:3). It was thus possible to assign the numbers 6, 8, 9, and 12 to the four fixed strings of the lyre, and to determine the intervals of the diatonic scale as 9:8, 9:8 and 256:243. From these observations and speculations the Pythagoreans built up … a deductive system of musical theory based on postulates or 'axioms' (a term that has a function similar to its use in mathematics). The dependence of musical intervals on mathematical ratios was thus established.

three to six, children from six to ten, and pupils from ten to eighteen or nineteen.[7] Up to the age of ten the curriculum was based on music (as a vocal art) and gymnastic, studies that were appropriated to the pre-rational formation of the child and that were intended to form the character and to cultivate bodily grace. The curriculum for the older pupils seems to have been divided into three sections of about three years each, with an emphasis on literature from ten to thirteen, music from thirteen to sixteen and mathematics from seventeen to about nineteen.

Intellectual studies were then interrupted by a period of gymnastic service intended to cultivate military preparedness, after which some students went on to higher education. For ten more years they studied mathematics at a higher level in order to understand the principles of the preparatory disciplines. Thus, by coordinating and combining the various branches of study, students gradually developed 'the faculty to detect the unity' behind the preparatory disciplines, their mutual relationships and the 'nature of the fundamental reality from which they all derived'.[8] When the course in higher mathematics was completed, a select number of students began the study of philosophy, the science of the good. The principal tool employed in this five-year course was dialectic, which included the rigorous task of reducing mathematics to logic. Emancipated from knowledge obtained by experience, the students of dialectic could finally arrive at rational truth.

The innovation in Plato's system of education was his emphasis on mathematics, an emphasis that was due to Pythagoras and his followers. Pythagoras, however, chose to reveal his teachings to his most advanced disciples only, while Plato argued that all men (but not all mankind) could learn mathematics, since men possessed the necessary requisite for that learning, namely, a reasoning faculty. Nevertheless, both mathematics and music began to drop out of the curriculum toward the end of the Hellenistic period; and by the Roman period the curriculum for training the elite was chiefly literary, oriented around the subjects of the trivium. This orientation

[7] For Plato's ideas about lower education, see Morrow, *Plato's Cretan City* (note 5), pp. 297–352.

[8] Marrou, *A History of Education* (note 5), p. 76.

continued throughout the Middle Ages into the seventeenth century, since literary subjects helped to illuminate study of the Bible, the principal focus of Christian education. Thus, classical instruction—study of the Latin language and literature—became the norm for lower education, which from about the twelfth century in England was increasingly centred in the grammar schools, whereas the chief centre of higher education was the universities, founded soon afterwards. The university curriculum was not classical but scholastic, emphasis being on the logical works of Aristotle and on disputation as the means of testing and establishing truth. Consequently, until about 1660 mental discipline was cultivated in the English universities by the study of Aristotelian logic—syllogistic reasoning—not mathematics. Hence, music and mathematics were not organised as part of the customary routine of instruction in English grammar schools or universities. At most, they were 'extras', available only on payment of extra fees.[9]

The classical and scholastic curriculum of conventional medieval and renaissance education was intended to educate the upper classes. Neglect of music and mathematics, therefore, was due to the specialist nature of those subjects, which had come to be

---

[9] For reasons why the grammar school played scarcely any appreciable part in the development of music or mathematics, see F. Watson, *The English Grammar Schools to 1666* (Cambridge, 1908), pp. 217, 221 and *The Beginning of the Teaching of Modern Subjects in England* (London, 1909), pp. xxii–xxiii. In the universities music as a vocal art was available only as an extra-curricular subject, even after the music professorships were established at Oxford in 1627 and at Cambridge in 1684. For the evidence, see N. C. Carpenter, *Music in Medieval and Renaissance Universities* (Normal, Oklahoma, 1958), pp. 76–90, 113–115, 153, 210, 284–294, 335–348; see also C. Wordsworth, *Social Life at the English Universities in the Eighteenth* Century (Cambridge, 1874) and *Scholae Academicae: Some Account of the Studies at the English Universities in the Eighteenth* Century (Cambridge, 1877). According to W. J. Galloway, *Musical England* (London, 1910), there was no requirement for residence as a necessary qualification for music degrees. Moreover, until the twentieth century the English universities were content to examine without systematically teaching music to intending degree candidates. Carpenter attempts to document that music as a mathematical art was included in the curriculum of English universities; whereas Foster, *The Beginning*, p. 254, claims that, even though the universities professed to teach subjects in the quadrivium, 'such teaching was evidently perfunctory, even when it existed, till the institution of Professors of Mathematics at Oxford in 1619 and at Cambridge in 1664'.

regarded more as mechanical, than as liberal, arts.[10] In ancient Greece the liberal and mechanical arts were so-called, because they constituted the study of the free man and the slave, respectively. Since the aristocracy set Greek standards of education, preference was for liberal, not mechanical, training. The ideal of liberal education, which was maintained throughout the Middle Ages and Renaissance, was in harmony with the interests of the upper classes—whether the free man of Greece, the cleric of the Middle Ages or the lay gentleman of the Renaissance. These men did not need to earn their living but could give time to subjects that had cultural value. Special skill, then, was the province of unfree men—those who had to earn their living—and technical or vocational training was not esteemed.

In 1622 Henry Peacham expressed the ancient point of view, when he wrote that the 'Mechanical Art and Artists, whosoever labour for their livelihood and gaine, have no share at all in Nobility and Gentry'.[11] Accordingly, the upper classes relegated the study of music and mathematics to leisure hours as recreational, not essential, parts of liberal education, since proficiency in any subject carried the taint of professionalism and implied time misspent.[12] Since they could afford to hire private tutors, the upper classes had the freedom and liberty to study recreational subjects not only as extras at the grammar

---

[10] The seventeenth-century mathematician, John Wallis, who had studied at the University of Cambridge, observed that mathematical subjects were 'scarce looked upon as *Academical* Studies, but rather *Mechanical*; as the business of *Traders, Merchants, Seamen, Carpenters, Surveyors of Lands,* or the like; and perhaps some *Almanack-makers in London*'. See W. T. Costellow, *The Scholastic Curriculum of Early Seventeenth-Century Cambridge* (Cambridge, Massachusetts, 1958, p. 103). See also J. C. Kassler, *The Science of Music in Britain, 1714–1830: A Catalogue of Writings, Lectures and Inventions* (2 vols., New York, 1979), vol. 2, pp. 1039–1042, for some of Wallis's contributions to the science of music.

[11] H. Peacham, *The Compleat Gentleman* [1634]...(Oxford, 1906), pp. 12–13.

[12] Peacham, ibid., warned young gentlemen that music could cause them to neglect their chief duties by making them addicted 'wholly to trifles'. Consequently, he recommended that gentlemen learn only 'to sing your part sure, and at first sight, withal, to play the same upon your Violl, or the exercise of the Lute, privately to yourselfe'. He also regarded mathematics as a recreational study.

school and university but also as part of private education at home, at courtly academies and inns of court, or abroad.

The method for training specialists in music and mathematics differed from that of amateurs, for about the same time that the grammar schools and universities were founded in England, the mechanical arts were incorporated into various guilds.[13] These guilds required artisans to be licensed in order to obtain the freedom of a particular town and the liberty to practise their craft in that town. Freedom, in legal language, meant special exemption from a general rule of compulsion; liberty implied a special privilege to do something forbidden to others.[14] To become a specialist in music or in mathematics, therefore, a child had to serve an apprenticeship, a term that denoted either the servitude of the apprentice or the duration of his indenture.

The duration of apprenticeship depended upon the nature of the subject studied or the locale in which a person was bound apprentice. At the end of the period specified, the apprentice was released from his indentures and granted freedom and liberty by the guild. Then, and only then, was he permitted to become a master, take apprentices and practise his craft.[15] Thus, the medieval and renaissance economic

---

[13] See *English Gilds: The Original Ordinances of More than 100 Early English Gilds*...(London, 1870). Some of the records of the principal London guild of musicians, the Worshipful Company of Musicians, are now deposited in the Guildhall Library, London. A few of these documents are printed in *The Worshipful Company of Musicians* (2d edn., London, 1905) and in H. A. F. Crewdson's *A Short History: The Worshipful Company of Musicians* (London, 1950, addenda and corrigenda, 1956). For the Company and the ancient Society of Waits, see also the pioneering study by W. L. Woodfill, *Musicians in English Society from Elizabeth to Charles I* (Princeton, 1953).

[14] See C. Wilson, *England's Apprenticeship, 1603–1763* (London, 1965).

[15] I. Pinchbeck and M. Hewitt, *Children in English Society*...(2 vols., London, 1969–73) detail conditions of apprenticeship; but they do not include any facts relating to indentures of musicians. See also 'Apprenticeship', *British Encyclopedia , or Dictionary of Arts and Sciences*...(6 vols., London, 1808–09), vol. 1, n.p., which states that during the Middle Ages the incorporated trades were called universities, the proper Latin name for any incorporation. Thus, there was a university of smiths, a university of tailors, etc. When the incorporations now peculiarly called 'universities' were first established, the term of years during which it was necessary to study in order to obtain the

system was protectionist, not free; and protection was sought chiefly from itinerant persons who went from house to house, from persons in other trades who practised music or mathematics as a sideline and from foreign competitors.[16] Protection from competition, however, was achieved not merely through guild regulation but also through preserving trade secrets.

Ancient Musical Doctrine

Because of the extra-curricular status of music and mathematics in the education of amateurs and because of the conditions of craft secrecy in the education of specialists, developments in music both as a vocal and as a mathematical art must be looked for outside traditional channels of education. The principal sources for these developments are systematic writings that treat the doctrine of music either as a speculative or as an applied science.[17] In the first kind of

---

degree of Master of Arts was taken over from the term of apprenticeship in the common trades—seven years. This duration was still required in the statute of apprenticeship drawn up during the time of Elizabeth I.

[16] On 24 April 1469 The Brotherhood of the King's Minstrels complained that

> certain ignorant rustics and craftsmen of various callings in our Kingdom of England have falsely represented themselves to be Minstrels. Of whom some assuming our licence which has in no way been granted to them, represent themselves as our Minstrels; under the colour of which licence and of the said art or occupation of Minstrels they in various parts of our Kingdom collect and receive large sums of money from our subjects.
>
> And although they are in that art or occupation by no means learned or skilled, and practise divers arts and callings on holidays and ordinary days and derive a sufficient living therefrom, they nevertheless move from place to place on festival days and collect all those profits by means of which our aforesaid Minstrels and our other Minstrels for the time being, sufficiently learned and instructed in the aforesaid art or occupation, and practising no other craft, trade or calling, should obtain their living.

See *The Worshipful Company of Musicians* (note 13), pp. 21–22, where the complaint and the Royal Charter granted to the Brotherhood are translated from the Latin into English.

[17] For the different classes of music textbooks, see Carpenter, *Music in Medieval and Renaissance Universities* (note 9), pp. 25–31; for the textbooks themselves,

science, the subject was world harmony, a speculative construct about the order of elements in the physical and ethical world. In the former world, where order was eternal and unchanging, the elements were the planets in their situation, motion and nature; the years in the succession of spring, summer, autumn and winter; the seasons in the alternation of day and night; and the months in the waxing and waning of the moon. In the latter world, where order was variable, the elements of the human body included the powers by which the human body grows, the mixture of the fluids or humours by which the body subsists and the mechanical activities of the body, whereas elements of the soul were such qualities as virtue and vice, joy and sorrow, pleasure and pain, and the bond between body and soul was created by the sympathetic relationships of all these elements for the purpose of imparting motion and sensation to the body.

In Plato's *Timaeus* world harmony appears as a musical harmony, inaccessible to human ears but comparable to human music, and since reducible to numbers to some degree accessible to human reason. Music, therefore, was a symbol as well as an imitation of the totality of the world, so that the speculative science of music became a key to understanding that world as well as a rationale for the practice of music, which could be explained as symbols or imitations of the cosmic order. This approach to music remained current for nearly 2,000 years in the West, for it continued ever-present in the writings of the Middle Ages and Renaissance.[18] Indeed, until about the end of the sixteenth century, the unquestioned authority on the musical doctrine of antiquity was Boethius, whose *De institutione musica*, set forth and defined the three divisions in the speculative

---

see the volumes in series B.III and B.VI of *Répertoire internationale des sources musicales*. For musical doctrine in other writings, e.g., poetic, alchemical, astrological, etc., see L. Thorndyke, *A History of Magic and Experimental Science* (8 vols., New York, 1923–58).

[18] L. Spitzer, *Classical and Christian Ideas of World Harmony: Prolegomena to the Word Stimmung* (Baltimore, 1963), who ranges widely through documents in different languages to demonstrate the currency of the Platonic ideas of world harmony; see also J. Hollander, *The Untuning of the Sky: Ideas of Music in English Poetry 1500–1700* (Princeton, 1961).

science of music: *musica mundana*, *musica humana* and *musica instrumentalis*.[19]

In addition to the speculative science of music, there was an ancillary and subordinate practical (applied) science. This second science was divided into two branches of study that originated in the development of the curriculum in antiquity. The first and introductory branch of study was music as a vocal art, knowledge of which was preliminary to the second more advanced branch, music as a mathematical art. This division retained the ancient association of music with the arts of the trivium and quadrivium, so that well into the seventeenth century, writers of treatises on the practical science of music employed two different approaches to their subject, both of which were technical in character.

On the one hand, some writers attempted to reduce music to elements in tone and time and to decide which of these elements and their combinations were suitable for use in musical performance and musical composition. These writers tended to duplicate the form and phraseology of definitions in linguistics and to represent music as a language of communication. Since this type of approach was concerned with the structural categories of musical grammar and musical rhetoric, we may call it 'philological'. On the other hand, some writers endeavoured to comprehend musical phenomena solely in terms of string or pipe lengths and to employ ratios to represent the sense distinctions of high and low pitch. These writers tended to duplicate the form and phraseology of definitions in arithmetic and geometry and to exhibit a conception of music as number and its proportions. Since this type of approach was concerned chiefly with the metrical, mechanical and acoustical aspects of music, we may call it 'mathematical'.[20]

Ancient and medieval lore held that a complete musician should possess qualifications in both branches of the practical science

---

[19] Spitzer, ibid., pp. 34–35.

[20] For the two traditions, see Lippman, 'The place of music' (note 3), who states that they may even be represented side by side in the same treatise. However, the duality of the sources is clearly reflected in the diverse definitions in treatises, in which the pratical science of music is defined as treating either 'proper movement, of relative quantity, or of measurement in relation to sound' or 'the science of poems and songs', 'the science of melody and singing'.

of music; that is, he should study music as a vocal and as a mathematical art. Thus, Guido of Arezzo, an eleventh-century theorist, distinguished the mere maker of music from the real musician by asserting:

> *Musicorum et cantorum magna est distantia,*
> *Isti dicunt, illi sciunt, que componit Musica.*
> *Non qui facit, quod no sapit, diffinitur bestia.*[21]

And this distinction was retained into the seventeenth century, for as late as 1653 we find a more detailed definition of a '*Complete Musitian*' as

> Him to be such, as hath not only Nibbled at, but swallowed the whole *Theory* of Musick; *i.e.* haveing profoundly speculated the *Pythagorean* Scheme of the various Sounds arising from various Hammers, beaten on an Anvill, respective to their different *Weights*, doth clearly and distinctly understand as well the *Arithmetical*, as *Geom[e]trical* Proportions of *Consonances*, and *Dissonances*: for, it is not the mere *Practical* Organist, that can deserve that Noble Attribute... .[22]

In addition to the theory of music, however, a complete musician required knowledge of 'all kinds of Humane Learning', for

> He must be a *Physiologist*; that He may demonstrate the Creation, Nature, Proprieties, and Effects of a Natural Sound. A *Philologer*, to inquire into the first Invention, Institution, and succe[e]ding Propagation of an Artificial Sound, or Musick. An *Arithmetician*, to be able to explaine the Causes of Motions Harmonical, by Numbers, and declare the Mysteries of the new Algebraical Musick. A *Geometrician*; to evince, in great variety, the Original of Intervalls Conson-dissonant, by the Geometrical, Algebraical, mechanical Division of a Monochord. A *Poet*; to conform his Thoughts, and Words to the Lawes of praecise Numbers, and distinguish the Euphonie of Vowells and Syllables. A *Mechanique*; to know the

---

[21] M. Gerbert, *Scriptores ecclesiastici de musica sacra potissimum*...(2 vols., San Blasianis, 1784), vol. 2, pp. 22–34.

[22] 'The stationer to the ingenious reader', *Renatus Des-Cartes Excellent Compendium of Musick*...(London, 1653), n.p.

> exquisite Structure or Fabrick of all Musical Instruments, Winde, Stringed, or Tympanous *alias* Pulsatile. A *Metallist*; to explore the different Contemperations of Barytonous and Oxytonous, or Grave and Acute toned Metals, in order to the Casting of tuneable Bells, for Chimes, &c. An *Anatomist*; to satisfie concerning the Manner, and Organs of the Sense of Hearing. A *Melothetick*; to lay down a demonstrative method for Composing, or Setting of all Tunes, and Ayres.

Last, but not least, the complete musician must be

> so far a *Magician*, as to excite Wonder, with reducing into Practice the Thaumaturgical, or admirable Secrets of Musick: I meane the Sympathies and Antipathies betwixt Consounds and Dissounds; the Medico-magical Virtues of harmonious Notes (instanced in the Cure of *Sauls* Melancholy fitts, and of the prodigious Venome of the Tarantula, *&c.*) the Creation of Echoes, whether Monophone, or Polyphone, *i.e.* single or Multiplied together with the Figure of Buildings, and arched Rocks, neer Rivers, Dales, or Woods, requisite to the multiplyed Reverberations of Sounds; the Artifice of Otocoustick Tubes, or Auriculary Meanders, for the strengthning, continuation, and remote transvection of weake sounds, and the mitigation of strong; the Model of Autophonous, or speaking Statues; and, finally, the Cryptological Musick, whereby the secret Conceptions of the mind may be, by the Language of inarticulate Sounds, communicated to a Friend, at a good distance.[23]

Both branches in the practical science of music were employed for practical purposes by mechanically as well as by liberally educated men, so that throughout the Middle Ages the terms 'practical' and 'speculative' designated two essentially separate spheres of human knowledge. The practical science of music was the elementary study for the man who intended to make music as well as the more advanced study for the man who wished to inquire into the nature of musical sound, whereas the speculative science of music was the study of the philosopher, the man who would explain the world. A more modern view began to make its appearance when writers spoke of the different branches in the practical science of music as being both theoretical and applied. For example, in 1531

---

[23] Ibid.

Juan Luis Vives, an important teacher of the early Renaissance, insisted that the arts of the quadrivium each had two aspects:

> the one which consists of the contemplative attitude is called theoretical (speculative); the other issues in work and is called practical (actuosus or effectrix). From the former the latter takes its source.[24]

As the science of music became increasingly delimited in the sixteenth century, there was a concomitant loss of feeling that music could symbolise the order of the universe; and the consequent 'demusicalisation' of western thought became evident in two ways. First, there was a dislocation of the semantic field, whereby the Platonic construct of world harmony, which once provided a unified conception of the world, was reduced to a mere rhetorical device.[25]

---

[24] F. Watson, *Vives on Education: A Translation of the De tradenis disciplinis...together with an Introduction* (Cambridge, 1913), p. 201. Vives visited England between 1523 and 1528, but some of his ideas were disseminated there and on the continent by Jan Amos Comenius, who laid down a method of teaching the arts, the first rule of which is an aphorism:

> 1. 'Theory,' says Vives, 'is easy and short, but has no result other than the gratification that it affords. Practice, on the other hand, is difficult and prolix, but is of immense utility.' Since this is so, we should diligently seek out a method by which the young may be easily led to the practical application of natural forces, which is to be found in the arts.

See J. A. Comenius, *The Great Didactic* [1627] tr. ...by M. W. Keatinge (London, 1910), p. 194. According to G. E. L. Owen, D. M. Balme, L. G. Wilson and L. Minio-Paluello, 'Aristotle (384–322 B.C.)', *Dictionary of Scientific Biography* ed. by C. C. Gillispie (16 vols., New York, 1970–80), vol. 1, pp. 250-281, Aristotle had distinguished between theory as contemplative and practice as active, so that after the transmission of Aristotle's ideas to the Latin Middle Ages, many writers continued to hold the Platonic conception of music as a speculative science alongside the Aristotelian' conception. For the latter conception to prevail, however, the ancient speculative science of music had to be rejected by consensus.

[25] Spitzer, *Classical and Christian Ideas* (note 18) and Hollander, *The Untuning of the Sky* (note 18), both treat the dislocation of the semantic field, although the latter author treats this dislocation in greater detail by showing the ways in which the construct, world harmony, was trivialised in poetry. By trivialisation Hollander means trifling—the reduction of *mundana* to mundane. However,

Second, there was a gradual disappearance of writings employing the ancient notions of world harmony as a principal mode of explanation. Indeed, Johannes Kepler was the last important writer on a quadrivial subject to employ the speculative science of music; but his *Harmonice mundi* (1619) is a Janus-faced work. In it, Kepler harks back more than two millenia to the Pythagorean music of the spheres at the same time that he signals the approach to a new mathematisation of science, for he quantified the third planetary law—that the squares of the periods of the planets are proportional to the cubes of their mean distances from the sun.[26] Thus, Platonic ideas of world harmony, along with the ancient speculative science of music, began to disappear just when modern science was emerging.

THE BEGINNINGS OF MODERNITY

The variety and diversity of subjects in modern science sprang from the mathematical arts of the quadrivium, so that, by the early part of

another meaning is implicit in his work, namely, the reduction of a quadrivial notion to a trivial one. At the same time that the dislocation of the semantic field occurs, there was a modification and eventual decline in use of the ancient mimetic doctrine that was part of the speculative science of music. The renunciation of the doctrine of imitation came first in relation to music and only later in relation to the other arts. See M. H. Abrams, *The Mirror and the Lamp: Romantic Theory and the Critical Tradition* (London, 1953); see also J. D. Boyd, *The Function of Mimesis and its Decline* (Cambridge, Massachusetts, 1968). See also Kristeller, 'The modern system' (note 2), p. 200; and, especially, E. Cassirer, *The Philosophy of the Enlightenment* tr. by F. C. A. Koelln and J. P. Pettegrove (Boston, 1965), who traces the change by examining the terms 'nature' and 'genius'. When the ancient doctrine of imitation was subverted, the notion of plagiarism makes its first appearance.

[26] E. Werner, 'The last Pythagorean musician: Johannes Kepler', *Aspects of Medieval and Renaissance Music*...(New York, 1966), pp. 867–882, who describes the modifications Kepler made to ancient Pythagorean doctrine. After 1619 the ancient science of speculative music does not appear in works of comparative stature to Kepler's on a quadrivial subject, although remnants of the tradition may be found in writings such as J. Webster, *Academiarum examen...offer'd to the Judgments of all Those that love the Proficience of Arts and Sciences and the Advancement of Learning*...(London, 1654), p. 42, where the author queried to what extent music may be useful to the sciences 'and the laying open of the universal Harmony of the whole Mundane Fabrick'.

the seventeenth century, mathematics not only included arithmetic, geometry, astronomy and music but also a good many other subjects besides.[27] But the extensiveness of subjects placed under the mathematics raises the question: what did writers of the latter part of the sixteenth and earlier part of the seventeenth centuries mean by the word?[28] The answer appears to be that the mathematics were so called, because they were learned from a master. This answer has its roots in the ancient world, when the great amount of training required to become skilled in any subject, and the need to have a master reveal the secrets of the arts, preserved the idea of mathematics as a mystery. Hence, the study of mathematics was allied to that of magic. The ancient Pythagorean doctrines of number as a key to the mysteries of all nature made mathematics a system of secrets that

---

[27] See, e.g., E. G. R. Taylor, *The Mathematical Practitioners of Tudor and Stuart England, 1714–1840* (Cambridge, 1966), who covers such subjects as surveying, cartography, navigation, sun-dial making and the like, although she completely overlooks subjects pertinent to musico-mathematical practitioners. That various crafts in music were regarded as mathematical is clearly shown in Hans Holbein's painting, 'The Ambassadors', reproduced in M. Boas, *The Scientific Renaissance 1450–1630* (London, 1962), where a detail in the painting depicts mathematical instruments of the early sixteenth century along with a lute and a music book. See also E. Winternitz, 'Leonardo da Vinci as a musician', *Times Literary Supplement* (1 April 1977), p. 411, who concluded that Leonardo's notes and designs relating to music were

> not merely diverting devices for performing magic tricks, but … [served] systematic efforts by Leonardo to realise some basic aims, of which the most important are the following: automation of certain instruments and facilitation of playing technique through new kinds of keyboards; increasing the speed of playing; extension of tonal range to make it possible, for instance, to play melodies on drums; overcoming the quick fading of the sound of plucked strings … by giving the instruments an endless bow; enriching comparatively simple instruments to make them capable of polyphony or a wide range of successive tones; and even having a polyphony of the sounds of bowed strings at the control of a keyboard.

See also Hunt, *Origins in Acoustics* (note 4), pp. 74–78, 125, 134, 144. Leonardo, however, was a complete musician, for he was an *Improvvisatore* or maker of songs, as well as a musico-mathematical practitioner.

[28] According to the *OED*, the Greek term for mathematics denoted something learned or doctrines involving geometrical reasoning, whereas in the Middle Ages the term 'mathematician' could mean astrologer.

required mastery reserved only for the initiated. In the later Middle Ages, when the crafts were incorporated into guilds, the various arts were referred to as 'mysteries', a term that was retained in the statute of apprenticeship drawn up during the time of Elizabeth I.[29]

The gradual change from craft mystery to physical science as the basis for technology was precipitated in the early part of the seventeenth century by the work of three intellectuals: Galileo Galilei, Francis Bacon and René Descartes. These men spoke against the emptiness of medieval philosophy and for the need to rebuild the structure of knowledge, although they differed in the means proposed or employed for discovering new knowledge. Until Galileo, physical science and ethics were part of philosophy, which—aided by the speculative science of music—was supposed to explain all relationships, physical and spiritual, quantitative and qualitative, in the world. The ancient speculative science of music set forth the Greek humanistic view of the world and man, for its doctrine identified 'what lofty purpose would reveal divine intelligence'.[30] In the seventeenth century, however, physics and ethics became, respectively, part of natural and moral philosophy.[31] Galileo provided

---

[29] For this statute, see Pinchbeck and Hewitt, *Children in English Society* (note 15). The Greek term for mystery denoted secret ceremonies, whereas Christian usage referred to religious truth known only by divine revelation. Mystery also came to mean a hidden or secret thing, a matter unexplained or inexplicable, as well as a handicraft, trade, art, skill or even guild. This last usage is found in the Byelaws of the Worshipful Company of Musicians, confirmed by the Lord Chancellor of England on 25 August 1606, and in the Act of Common Council of 11 September 1700 relating to the same Company; see *The Worshipful Company of Musicians* (note 13), pp. 47, 83, 85, 88.

[30] C. C. Gillispie, *The Edge of Objectivity: An Essay in the History of Scientific Ideas* (Princeton, 1967), p. 14.

[31] Physics, though not ethics, ultimately became a separate discipline from philosophy; but the religious motive in science remained strong during the seventeenth century, and J. A. Comenius was not singular in issuing a work with the title, *Natural Philosophie Reformed by Divine Light: or, a Synopsis of Physicks...Exposed to the Censure of Those that are Lovers of Learning, and Desire to be Taught of God*...(London, 1651). By the eighteenth century, however, explanation was becoming increasingly secular, so that some clergymen attempted to reverse the trend. For one such clergyman, see Chapter 8 *infra*.

the impetus for this separation of philosophy by maintaining there were two qualities in bodies: primary and secondary. He defined primary qualities as properties essential to metrical description and secondary qualities as modes of perception in humans. In *The Assayer* (1623), for example, he stated that

> whenever I conceive any material or corporeal substance, I immediately feel the need to think of it as bounded, and as having this or that shape; as being large or small in relation to other things, and in some specific place at any given time; as being in motion or at rest; as touching or not touching some other body; and as being one in number, or few, or many. From these conditions I cannot separate such a substance by any stretch of my imagination. But what it must be white or red, bitter or sweet, noisy or silent, and of sweet or foul odor, my mind does not feel compelled to bring in as necessary accompaniments. Without the senses as our guides, reason or imagination unaided would probably never arrive at qualities like these. Hence I think that tastes, odors, colors, and so on are no more than mere names so far as the object in which we place them is concerned, and that they reside only in the consciousness.[32]

But it was on the basis of his musical experiments that Galileo formulated a new method of inquiry into primary qualities of bodies. According to this method, which he exemplified in *Discorsi e dimostrazioni matematiche, intorno a due nuove scienze* (1638), at each stage of scientific inquiry sense experiments must be combined with reasoning and with mathematics to afford a well-founded basis of deduction. Thus, began a movement away from traditional humanism toward a more naturalistic interpretation of the physical world.[33]

---

[32] S. Drake, *Discoveries and Opinions of Galileo...*(New York, 1957), p. 274. The distinction between primary and secondary qualities in bodies, but not the terminology, is also found elsewhere in Galileo's writings.

[33] S. Drake, 'Renaissance music and experimental science', *Journal of the History of Ideas* (1970), 31: 483–500, suggests that the modern experimental method in physics originates with the acoustical experiments of Giovanni Battista Benedetti, Vincenzio Galilei and his son, Galileo; see also C. Palisca, 'Scientific empiricism and musical thought', *Seventeenth Century Science and the Arts* ed. by H. H. Rhys (Princeton, 1961), pp. 91–137, and Kassler, *The Science of Music* (note 10), vol. 1, pp. 367–369.

Bacon's influence for the 'advancement of learning' rests chiefly upon his philosophy of scientific method, which has been termed 'inductivism' and which, beginning with facts rather than with principles, rises by successive steps to general laws. But Bacon insisted that the study of nature should proceed not only by induction but also by experiment.[34] Indeed, his emphasis on experiment shaped the course of later scientific investigation, so that 'the term "experimental science" has become practically a synonym for "modern science," and nothing so clearly differentiates post-seventeenth-century science from that of the Renaissance, or of Greece, as the role of experiment'.[35] Although experiment had been performed before Bacon, the traditional means for testing and establishing truth was the syllogism not the experiment or the mathematical equation.[36] Bacon attempted to repudiate syllogistic logic by calling for experiments to be conducted that would determine how nature behaves, where questions posed and answers provided would arise from careful and extensive observation. In the conduct of experiments, therefore, he sought to minimise all hindrances to open-minded inquiry, including such things as tradition, training and philosophical systems. In stressing the inductive and empirical study of nature, Bacon helped to create a climate of opinion that discouraged speculation 'in favor of accumulation of detail, abstract generalisation in favor of natural history', for he disdained

---

[34] Bacon outlined his scientific method in two works: first, *The Twoo Bookes of the Proficience and Advancement of Learning* (1605)—expanded in the Latin version, *De augmentis scientiarum* (1623); then, the incomplete *Novum organum* (1620), published in parts. One of the parts, *Sylva sylvarum* (1626), treated music under the doctrine of sound and hearing and included numerous 'experiments' that were mostly pronouncements on sound and music; see Kassler, *The Science of Music* (note 10), vol. 1, pp. 37–39. For the role that tabulation of information played in Bacon's inductive method, see M. Hesse, 'Francis Bacon's philosophy of science', *Essential Articles for the Study of Francis Bacon* ed. by B. Vicars (Hamden, Connecticut, 1968), pp. 114–139.

[35] Gillispie, *The Edge of Objectivity* (note 30), p. 79.

[36] For a history of experiments before Bacon, see A. C. Crombie, *Robert Grosseteste and the Origins of Experimental Science 1100–1700* (Oxford, 1953), and L. Thorndike, *A History of Magic* (note 17). For the struggle to replace syllogistic logic with the logic of induction, see W. S. Howell, *Eighteenth-Century British Logic and Rhetoric* (Princeton, 1971).

anyone 'who would extend a few ideas or calculations into a system of the world'.[37]

In contrast to Bacon, Descartes sought to build a system of the world by a method that offered the same guarantee of accuracy, effectiveness and simplicity as algebra and geometry. Employed first in his *Musicæ compendium* and described afterwards in his *Discours de la méthode* (1637), Descartes proposed to embrace the world 'in the clear and simple, and therefore true, ideas of geometry, not like Plato by denying existence to the merely physical, but by a total criticism that would strip philosophy of the illusion that understanding may be had on vaguer terms'.[38] Descartes, therefore, called for a reconstruction of knowledge out of the mathematical concepts of number, motion and extension. To assist in this reconstruction, he invented analytical geometry by applying algebra to spatial relationships. By abolishing the distinction between algebra as the mathematics of discrete quantity and geometry as the mathematics of the spatial continuum, Descartes put in place a method that, in principle, involved no concept that could not be explained in terms of pure geometry together with the dimension of time.[39] Unlike Galileo, who saw mathematics as a tool of science, a means of expressing quantity, Descartes sought to explain the behaviour and reason of all natural things by mechanistic concepts, thereby overestimating the function of mathematics in scientific explanation. Nevertheless, by showing the purely mechanical nature of physical processes, Descartes paved the way for the modern application of mathematical methods to all science.

The work of Galileo, Bacon and Descartes was furthered in the British Isles by a Puritan reform movement, which saw education as

---

[37] Gillispie, *The Edge of Objectivity* (note 30), p. 77.

[38] Ibid., p. 86. Descartes wrote his music treatise in 1618; it was published in Latin in 1650; an English translation appeared in 1653 (see note 22). B. Augst, 'Descartes's Compendium of Music', *Journal of the History of Ideas* (1962) 26: 119–132, describes Descartes' attempt to reduce sensory perception to a strictly mechanical and, therefore, measurable function. Hollander, *The Untuning of the Sky* (note 18), pp. 177–179, suggests that Descartes' treatise contains an early statement of the doctrine of affections.

[39] Gillispie, *The Edge of Objectivity* (note 30), pp. 87–88.

the key to improving religion and society. The Puritan reformers called for the introduction of new subjects, chiefly scientific studies of nature and human society. Instead of liberal education, they valued technical training as the means to material betterment, for they held that a man should be educated for a specific purpose and not just educated. This purpose, however, was to be determined by the sphere of life in which a man was to act, as well as by his natural capacities. Thus, even though the reformers esteemed special skills, they still held that the province of specialisation should be class-designated. Despite this limitation in their programme of reform, the Puritans, and particularly those in the circle around Samuel Hartlib, gave the advancement of learning a central role by

> contributing schemes covering the reorganisation of all levels of the educational system, from poor school to university. They proposed a reorganisation of the curricula, outlined new pedagogical methods and produced new textbooks with an explicit psychological foundation. ... For the first time in English history, concentrated efforts were made to outline detailed models of a state controlled comprehensive educational system. Parliament was now expected to play a dominant role in educational policy, a function which it had previously not had the opportunity to exercise.[40]

Although Parliament and the conservative legislators of the interregnum prevented most of the reforms from being implemented, the Puritans' emphasis on 'real' studies—that is, the mathematical arts of the quadrivium—helped to subvert the classical and scholastic curriculum of conventional liberal education.[41]

---

[40] C. Webster, *Samuel Hartlib and the Advancement of Learning* (Cambridge, 1970), pp. 4–5; see also Webster's *The Great Instauration: Science, Medicine and Reform 1626–1660* (London, 1976), and R. L. Greaves, *The Puritan Revolution and Educational Thought* (Newark, New Jersey, 1969). After publication of Bacon's *Advancement of Learning* (note 34), the phrase appears in many titles throughout the seventeenth century, especially in the writings of the reformers. See, for example, Webster, *Academiarum examen* (note 26), and T. Salmon, *Proposal to the Advancement of Musick*...(London, 1672), whose plan to reform musical notation took its cue from Bacon's work.

[41] In 1656/7 a petition was made by the musician, John Hingston, and others to 'The Right Honourable Committee of the Councell of the Advancement of Musicke' to establish a 'College' of musicians; see J. Pulver, 'A musical

Changes began to be effected even earlier in the traditional channels of mechanical training, one of the most innovative reforms being Gresham College, which provided the first endowed chairs for both music and mathematics. Founded in 1575 by Sir Thomas Gresham and opened in 1597, the College offered lectures on seven subjects: rhetoric, medicine, law, geometry, music, astronomy and divinity. The musician, Thomas Ravenscroft, not only expressed gratitude to Gresham for opening a new educational centre for the artisans of London but also credited him with reviving the ancient 'liberal Arts and Sciences, especially the Mathematicks, which were somewhat neglected euen in the Universities'.[42]

In the seventeenth century Oliver Cromwell allowed demobilised soldiers into protected trades without apprenticeship; and after the great fire of London in 1666 the shortage of artisans dealt the craft system another blow. Moreover, changing notions of freedom and liberty challenged guild regulation, so that by the eighteenth century attempts were made to repeal the statute regulating apprenticeships, which increasingly was regarded as a monopoly and against competition.[43] In addition, the remarkable degree of social mobility in the seventeenth and eighteenth centuries, especially

---

petition', *Musical Opinion* (1930), 53: 809, and P. Scholes, *The Puritans and Music*...(London, 1934), pp. 282–285. Hingston's petition was followed by a number of others, all unsuccessful until 1822 when a national music school was founded to train professional musicians of both sexes; see J. C. Kassler, 'Burney's *Sketch of a Plan for a Public Music-School*', *The Musical Quarterly* (1972), 58: 210–234.

[42] T. Ravenscroft, *A Brief Discovrse of the True (but Neglected) Use of Charact'ring the Degrees of their Perfection, Imperfection, and Diminution in Measurable Musicke, against the Common Practise and Custome of these Times*...(London, 1614). See also J. W. Burgon, *The Life and Times of Sir Thomas Gresham*...(2 vols., London, c.1850).

[43] Wilson, *England's Apprenticeship* (note 14), p. 136; see also *The British Encyclopaedia* (note 15), vol. 1, n.p., in which the legislation relating to apprenticeship is surveyed, and Crewdson, *A Short History* (note 13), p. 43, who notes that the last important action taken by the Worshipful Company of Musicians was in 1763 and that after that date 'there is no trace of any further attempt to maintain the Company's ancient powers, with the exception of a feeble gesture made in 1801'. By the nineteenth century, opinion was that trade 'should be as free as the air we breathe'.

between the middle and top ranges of society, also began to affect the power of guilds.[44]

The new utilitarian value system of the Puritans, the weakened power of craft guilds, the development of geometry, the increasing subordination of mathematics to experience and the discovery of unexpected laws through close observation—all these contributed to the era of modern science, which commenced with the science of Isaac Newton at the end of the seventeenth century and continues on to the present day. But the denomination 'modern science' refers not merely to natural science but to an era when the methods of science began to pervade all areas of human thought and endeavour. These methods helped to bring about a decline of craft mysteries, so that such pseudo-sciences as astrology, alchemy, cheiromancy and the speculative science of music began to fall away from mathematics 'into their own limbo, or attempted to join themselves to less exacting divisions of knowledge, till standards of proof and demonstration placed them outside the scope of scientific understanding'.[45] At the same time, there was an open publication of musical and other discoveries that was in direct contrast to the tradition of secrecy and concealment that had been necessary for the protection of medieval craftsmen.[46] These discoveries were made by non-initiated amateurs

---

[44] A. R. Wagner, *English Genealogy* (2d edn., Oxford, 1973), sorts out the shifts of meaning of class descriptions at different times and places, although he does not include musicians in his survey.

[45] Watson, *The Beginning* (note 9), p. 280, also states that until about the middle of the seventeenth century, it is not always easy to differentiate cabbalistical and astrological works from mathematical books, in which 'secrets' of all kinds were sometimes included. However, 'when the conception became general that mathematics were an accurate and profitable study for the evidence and certainty of their demonstrations, then the province of mathematics became delimited'. See also J. Ferguson, *Bibliographical Notes on Histories of Inventions and Books of Secrets* (2 vols., London, c.1959); and E. Wellesz, 'Music in the treatises of Greek gnostics and alchemists', *Ambix* (1951), 4: 145–158, whose study includes the alchemical treatise, 'On the making of gold', a book of secrets in which allusions to music are 'kept obscure in the same way as the descriptions of the alchemical process itself'.

[46] Although the first English patent for a musical invention is dated 1694, many discoveries and inventions were publicised before that date through the *Philosophical Transactions of the Royal Society*, through lectures to various

as well as by specialists, whose researches in the seventeenth century mark the beginning of a proliferation of writings on music.

MODERN MUSICAL DOCTRINE

The traditional vehicle for preserving and transmitting information about music had been the encyclopaedia, which was formed of separate textbooks on the seven arts.[47] During most of the Middle Ages, encyclopaedias were written in Latin, the international language of scholarly communication and instruction. The Reformation, coupled with the claims of great literature in the vernacular, undermined the authority of classical writers and led to the gradual replacement of Latin by English and other vernaculars. Thus, a broader public could be informed and enlightened.[48] The utility of English is demonstrated by the public to which music textbooks in the vernacular were addressed, for the audience was no longer limited to mechanically-educated specialists or liberally-educated amateurs but included untutored parishioners who wished to sing psalms by book.[49]

---

other learned societies and through treatises on specific subjects. For some of these discoveries and inventions, see Kassler, *The Science of Music* (note 10).

[47] Martianus Capella gave the definitive form of the encyclopaedia as seven arts or sciences; see W. H. Stahl, R. Johnson and E. L. Burge, *The Quadrivium of Martianus Capella*...(New York, 1970), who also include a bibliography of the seven arts in medieval and renaissance iconography.

[48] See R. F. Jones, *The Triumph of the English Language* (London, 1953).

[49] See, e.g., the twelve-page 'short Introduction into the Science of Musicke, made for such as are desirous to haue the knowledge thereof for the singing of ... Psalmes'. This treatise was included in various versions of the *Psalmes of David in Englishe Metre*, issued first in 1561. In the 1580 version the anonymous author stated that

> the rude and ignorant in song, may with more delight, desire, & good wil[l], be moued and drawne to the Godly exercise of singing of psalms, as well in common places of prayer, where altogether in one voice, render thanks and praises to God, as priuately by themselves, or at home in their houses. I have set here in the beginning of this book of Psalmes, an easy & most playne way and rule, of the order of the Notes and Kayes of singing which commonly is called the Scale of Musicke, or the

For the literate, upper-class amateur, however, the landmark music textbook is Thomas Morley's *A Plaine and Easie Introdvction to Practicall Mvsicke* (1597), the title-page of which is illustrated on the facing page.[50] This work is not the first music textbook in English, but its literary achievement was such that by the seventeenth century, the best music textbooks by English-speaking authors were published in the native tongue. The few works in Latin that continued to be issued were remnants of a dying tradition; and by the eighteenth-century Latin textbooks in music were quite anachronistic,[51] as was the encyclopaedia in its ancient format.[52]

Along with the proliferation of writings on music, there was a new treatment of its subjects, so that movement was toward a new conception of musical science, emerging from research in the ancient

> Gamma vt, Whereby (and diligence geuen thereunto) every man may in a few dayes, yea in a fewe houres, easily without all payne, & that also without ayd or help of any other teacher, attayne to a sufficient knowledge to sing any psalms cōtayned in this book or any such other playne & easy songes as these are.

See R. Steele, *The Earliest English Music Printing*...(London, 1903), who gives the entire 'short Introduction' in facsimile. The provision of information for self-instruction, of course, was contrary to the medieval notion of keeping trade secrets.

[50] See Kassler, *The Science of Music* (note 10), vol. 2, pp. 784–786. Like Bacon's phrase, 'advancement of learning', the phrase 'plain and easy' became a common theme in music textbooks, a theme already present in the 'short Introduction' quoted above (note 49) and repeated in such works as R. Barber's *Psalm Singer's Choice Companion: or a Plaine and Easy Introduction to Musick*...(London, 1727).

[51] Latin continued to be utilised for inaugural dissertations in medicine; see, e.g., I. Brown, *Disquisitio medica inauguralis, de sonorum modulatorum VI in corpora humana*...(Edinburgh, 1751), and L. Odier, *Epistola physiologica inauguralis de elementariis musicae sensationibus*...(Edinburgh, 1770).

[52] Among the last encyclopaedias to be issued in the old format were J. Newton, *English Academy: or a Brief Introduction to the Seven Liberal Arts*...(London, 1677); R. Blome, *Gentleman's Recreation*...(London, 1686, 2d edn., 1710); and H. Curson, *The Theory of Sciences Illustrated: or the Grounds and Principles of the Seven Liberal Arts*...(London, 1702).

VIRESCIT VVLNERE VERITAS
Ptolomeus
Marinus
Strabo
Aratus
Polibius
Hipparchus
Astronomia
Geometria
Musica
Arithmetica
MERCVRIVS
IB F
A
PLAINE AND
EASIE INTRODVCTI-
ON TO PRACTICALL
MVSICKE,
set downe in forme of a dialogue
Deuided into three partes,
The first teacheth to sing with all
things necessary for the knowledge of
pricksong.
The second treateth of descante
and to sing two parts in one vpon a plainsong or
ground, with other things necessary
for a descanter.
The third and last part entreateth of com
position of three. foure, fiue or more parts with
many profitable rules to that effect.
With new songs of. 2. 3. 4. and. 5 parts.
By Thomas Morley, Batcheler of musick, &
one of the gent. of hir Maiesties Royall Chappell.
Imprinted at London by Peter Short dwelling on
Breedstreet hill at the signe of the Starre. 1597.

practical science of music.[53] Neither the philological nor mathematical branches of this science were theoretical in the modern sense of the term; that is, they did not derive their principles from, nor demonstrate these principles by, the new mathematical and experimental methods. Instead, writers on music as a vocal art had defined their subject matter by means of Aristotelian categories, so that the various divisions and subdivisions of a topic were merely enumerated and referred to categorically as unchanging musical entities; whereas writers on music as a mathematical art had adopted 'number-theoretical conclusions that corresponded in simple cases to string-lengths and extended those incorrectly to tensions, weights and volumes'. In neither case did writers base their theories on actual sounds nor proceed by 'careful measurement, hypothesis, geometrical deduction, and test thereof'.[54]

In the early modern period of the seventeenth, eighteenth and first part of the nineteenth centuries, the dominant character of British musical science also was empirical, not theoretical, since the majority of writers were concerned merely to observe and catalogue musical elements by identifying and enumerating various sized particles of sound, intervals of the scale of music, structures in musical

[53] In writings on music as a vocal art, there is evidence of a transition from the medieval system of modality toward formulation of the modern system of tonality. As part of the change from the old system to the new, we find a gradual movement from polyphonic to homophonic musical thinking, in which the sonority of vertical harmonic structures gains prominence over the coincidence of intervals in simultaneously-sounding melody lines. Some of the steps leading to the formulation of modern harmonic theory and the modern system of tonality are described by W. T. Atcherson, 'National predilections in seventeenth-century music theory: England', and 'Key and mode in seventeenth-century music theory books', *Journal of Music Theory* (1972), 16: 6–15, (1973), 17: 204–232; L. F. Chenette, *Music Theory in the British Isles during the Enlightenment* (Ohio State University doctoral dissertation, 1967); H. Riemann, *Geschichte der Musik Theorie im 9.–19. Jahrhunderts* (Berlin, 1920); and M. Shirlaw, *The Theory of Harmony*...(2d edn., DeKalb, Illinois, 1955). For developments in music as a mathematical art, see Hunt, *Origins in Acoustics* (note 4).

[54] S. Drake, 'Ptolemy, Galileo and scientific method', *Studies in the History and Philosophy of Science* (1978), 9: 99–115, who also states that when 'string tensions were actually measured by Galileo's father [Vincenzio] they were found to contradict such theories, and a revolution in music swiftly followed'.

composition and simple or complex emotions aroused when experiencing music. These concerns did lead to an accumulation of new musical elements, but they did not lead to a new conception of musical science, for the observation of elements, without preconceived theory, is only natural history.

As writers on music continued to amass a vast store of disconnected but highly suggestive data, what was needed for musical inquiry were theories of music capable of coordinating existing data and, by their very generality, of orienting future investigation and discovery. Only a very few writers in the British Isles undertook to construct such theories, and those who did tended to explain the relationship of various musical elements one to another.[55] Their concern was with causality, that is, with the origins of the different effects observed in music, whether viewed as musical sound, musical arithmetic, musical structure or emotional experience. But they contributed to a new, mechanistic conception of musical science by utilising a description of music aided by mathematical concepts drawn from one or another type of seventeenth-century mechanics—for example, statics, kinematics or, after Newton, dynamics.[56]

Mechanistic thought stems in part from the application of mathematics to physical science from the end of the sixteenth century onwards, when Galileo insisted on the doctrine of relativity of motion and began an intellectual movement in which the investigation of natural laws as mathematical relationships was to replace traditional inquiries into the nature and essences of physical entities. Traditional inquiries had been Aristotelian, and the goal of Aristotelian physics was to define the essential nature of, and the proper place for, every species according to the purpose that it served. In Aristotelian physics elements were not thought of as a constituent part, a final result after

---

[55] E.g., syntax became a principal concern of writers on the science of music as a vocal art.

[56] For one example, see Chapter 8 *infra*; for other examples, see Kassler, *The Science of Music* (note 10). For the application of mathematical methods to physical science, see R. E. Schofield, *Mechanism and Materialism: British Natural Philosophy in an Age of Reason* (Princeton, 1970), and R. S. Westfall, *The Construction of Modern Science: Mechanisms and Mechanics* (New York, 1971). For historical background, see E. J. Dijksterhuis, *The Mechanization of the World Picture* tr. by C. D. Dijkshoorn (Oxford, 1969).

chemical analysis had done its work. Instead, matter and form were given a joint role, and

> the elements far from being ultimate and different indivisibles were primarily certain qualities attributable to all matter. They were founded on the notions of hot, cold, cry and moist ... [that is] the elements were thought of through their effects. These effects working on a common substance were thought, in co-operation with stellar influence and the occasional extraordinary intervention of God, to explain the way the sublunary world was conducted. ... The elements therefore as well as being effects were at least aspects of the common substance, and as such they had their almost ceremonial places in the great world order.[57]

In contrast to this approach, early modern writers outlined theories based on the hypothesis that music is a particulate mass, that is, it consists of differentiated classes of elements. They described these elements in terms of such geometrically definable quantities as size, figure, number and motion. Categories of explanation, however, depended upon the hypotheses underlying a theory. If the hypotheses stemmed from Cartesian mechanics, then the categories were number,

---

[57] E. M. W. Tillyard, *The Elizabethan World Picture* (London, 1943), p. 56; see also Owen, Balme, Wilson and Minio-Paluello, 'Aristotle' (note 24), pp. 255, 261, who state that Aristotle accepted from Plato's *Timaeus* the four elements (fire, air, water and earth)

> as combinations of hot, cold, wet, and dry: fire is hot plus dry, air is hot plus wet, and so on. In his system hot, cold, wet, and dry are the primitive qualities of matter, but cannot exist in isolation. Fire, air, water, and earth are the simplest separable bodies, and are transformable into each other.

This approach, which survived through the Middle Ages, means that the question, 'What are the essential attributes of matter?', must go unanswered, because Aristotle rejected all theories that explained physical change by the 'rearrangement of some basic stuff or stuff's', like atoms 'endowed with fixed characteristics' such as size, figure, number and motion. Indeed, he rebutted atomism in particular, 'arguing that movement in a vacuum is impossible … and that the concept of an extended indivisible body is mathematically indefensible'. In the British Isles the work of Robert Boyle and others in the seventeenth century helped to repudiate Aristotelian science and to gain acceptance of the new particulate theories of matter.

motion and extension; but if they derived from Newtonian mechanics, then the categories were mass, length and time.[58]

Whatever the hypotheses employed, writers of theories attempted to determine the scope of musical science. As this task began to be undertaken, two areas of musical knowledge were delimited, thereby bringing about a separation of its two ancient branches—philological and mathematical. By 1817, therefore, George Farquahar Graham could assert that the science of music consisted of

> I. That which regards the generation and propagation of musical sounds,—the particular qualities of musical sounds, and the ratios of commensurable relations which exist between those sounds in their comparative degrees of acuteness and gravity. This physico-mathematical branch of music falls properly within the province of the natural philosopher, mathematician, and algebraist, and belongs in strictness to the science of acoustics.
> II. That which regards the sensible phenomena of these sounds, and their mutual relations in melody and harmony, when brought under the general rules or principles of musical composition. This branch immediately concerns the musical critic, or the musical composer.[59]

Although the separation of the two branches did not begin until the end of the seventeenth century, it was the direct result of the estrangement of natural and moral philosophy arising from the work of Galileo in the early part of that century. I have already noted that

---

[58] In the eighteenth century English music theorists tended to favour Cartesian hypotheses; but the German-born musician and astronomer, Frederick William Herschel, adopted Newtonian hypotheses in his manuscript notes for a treatise on musical composition; see Kassler, *The Science of Music* (note 10), vol. 1, pp. 505–507.

[59] [G. F. Graham], *General Observations upon Music*...(2d edn., Edinburgh, 1817), pp. 16–17. So-called 'natural law' theories of musical composition, in which rules were deduced from the overtones of a vibrating string or from other natural causes, continued to unite the two branches of the practical science of music, while striking out in new directions by employing new analytical methods. By the second half of the eighteenth century, however, this type of theory construction was falling into disrepute, as, e.g., in responses to Jean Le Rond D'Alembert's commentary on the theory of Jean-Philippe Rameau, for which see Kassler, *The Science of Music* (note 10), vol. 1, pp. 249–253.

Galileo reduced physical nature to primary qualities that were objectively measurable and treated other aspects as secondary and subjective. This approach is also exhibited in the science of music in its early modern form: the mathematical branch took as its field primary qualities of sounding bodies that were objectively measurable, employed physics as its tool and fell under the domain of natural philosophy, whereas the philological branch took as its field secondary qualities of musical elements, defined these subjectively (where measurement is probable, not certain), employed linguistic analysis as its tool and fell under the domain of moral philosophy.[60]

Research in the mathematical branch led eventually to the modern science of musical acoustics, which does not occupy a particularly significant position in modern systems of knowledge or education.[61] Indeed, in its early modern form, the science of musical acoustics was treated variously as part of one or another subject in natural philosophy, being classed as a mixed mathematics under phonics, dynamics, pneumatics, hydrodynamics or physics. Its forerunner, the mathematical branch of the practical science of music, had provided the pupil with musical erudition and had made the difference between a mere maker of music and a complete musician.[62] But the modern science of musical acoustics came to be regarded as the province of the scientist, a term coined in 1840 by

---

[60] Measurement of secondary qualities of musical elements pertains in part to how various configurations of elements were determined as units of musical structure. For example, the musical phrase was a commonly accepted unit, but there was little agreement about the standard measurement for the size of this unit.

[61] Although some seventeenth-century musical scientists in the British Isles made important preliminary contributions to musical acoustics, the first modern textbook on the subject was by a German, E. F. F. Chladni; see Hunt, *Origins in Acoustics* (note 4), pp. 3–4.

[62] Remnants of this ancient notion may be found in the works of many writers. See, e.g., J. Robison, 'Temperament of the scale of music', *Supplement to the Third Edition of the Encyclopaedia Britannica*...(2 vols., Edinburgh, 1799–1801), vol. 2, p. 641, who asserted that to be able to tune a harpsichord 'with certainty and accuracy seems an indispensable qualification of any person worthy of the name of a *musician*'. In Robison's day, however, keyboard instruments were more often tuned by 'mechanics' than by musicians.

William Whewell to denote one who studies natural science, rather than one who is erudite, as the medieval term '*scientia*' implied.[63]

The philological branch of the practical science of music also underwent modification, and the modern science to emerge from this branch was the science of musical composition and criticism.[64] This new science was intended to equip a person to judge correctly the accuracy or regularity of grammatical and rhetorical facets of musical composition or to compose correctly himself. By the middle of the eighteenth century, however, the science of musical composition and criticism was considered of too great magnitude and intricacy for the grasp of immature or female intellects: the exercise of a vigorous and discerning mind was requisite to the acquirement of profound theoretical knowledge and the attainment of excellence in this field.[65] Precocious talent, such as Mozart's, came to be recognised; but evidence of such a talent was considered as rare as the discovery of a new planet.[66]

Thus, the new science of musical composition and criticism replaced the old mathematical branch as the advanced study for musicians, professional and amateur alike, for it was separated out also from the art of performance, which came to be regarded as an elementary study and as a necessary prerequisite for those who would

---

[63] W. Whewell, *History of Scientific Ideas: Being the first Part of the Philosophy of the Inductive Sciences* (3d edn., 2 vols., London, 1858), vol. 1, p. 113.

[64] For some contributors to this branch, see Kassler, *The Science of Music* (note 10).

[65] For women, see Kassler, 'Burney's *Sketch*' (note 42); for children, see J. C. Kassler, 'Music made easy to infant capacity 1714–1830', *Studies in Music* (1976) 10: 67–78. The notion that children and women were incapable of composing music or studying musical composition stemmed from contemporary doctrines of developmental psychology and education, most of which were indebted to the theories of John Locke. These doctrines were not challenged until nearly the second decade of the nineteenth century. See, e.g., Kassler, *The Science of Music* (note 10), vol. 2, pp. 720–723.

[66] Writings on musical 'prodigies' date from the second half of the eighteenth century; see Kassler, ibid., vol. 1, pp. 52–54, 132–133 and Chapter 10 *infra*.

compose.[67] At the same time, a breach was opened between the technical study of musical composition, on the one hand, and the science of musical criticism, on the other hand; and the first hints of romanticism occur as the latter science moved away from setting standards of musical excellence toward explaining only the emotional effects of music.[68]

Conclusion

The separation of the two branches of the science of music into subdivisions of natural and moral philosophy marks the beginning of the separation of art and science into 'two cultures'.[69] As a result, there was a gradual shift in meaning of the terms 'art' and 'science', a shift that coincided with the transition from the ancient conception of the encyclopaedia as seven arts to the modern conception, emerging in the eighteenth century, of universality of encyclopaedic knowledge in the arts, sciences and literature:[70] As the various departments of

---

[67] This development is traceable in part through treatises on thorough bass, the first of which were published in the seventeenth century; see F. T. Arnold, *The Art of Accompaniment from a Thorough-bass as Practised in the 17th and 18th Centuries* (London, 1961).

[68] Hollander, *The Untuning of the Sky* (note 18), p. 380, is partially correct in stating that the sharp division between amateur and professional musician drove a wedge between the critical memoir and the technical study of composition. But some philosophers—for example, David Hartley, in his *Observations on Man* (1749)—contributed to the separation of criticism from composition, because they tended to reduce music to emotional experience and to ignore the structural properties of music. Earlier, Roger North had recognised and, later, Thomas Reid was to recognise the problem such a reductionist approach posed for understanding music. For these philosophers, see Kassler, *The Science of Music* (note 10), vol. 1, pp. 30–34, 156–159, vol. 2, pp. 798–802, 873–876.

[69] C. P. Snow, *The Two Cultures…An Expanded Version…* (Cambridge, 1965), distinguishes chiefly between literature and science, but the 'cultures' demarcated from the eighteenth century were more than two, as may be seen from the titles of early modern encyclopaedias, for which see Kassler, *The Science of Music* (note 10), vol. 2, Appendix A.

[70] For the transitional period, see R. F. Young, *Comenius in England: The Visit…to London in 1641–1642; its Bearings on the Origins of the Royal*

knowledge within the field of music were subsumed under one or another subject in natural or moral philosophy, the province of art narrowed, for it came to designate the useful or the polite arts, thereby bringing about a distinction between craft and fine art.[71] When the modern system of the arts was codified at the end of the eighteenth century, therefore, only the fine art of music in performance or composition was admitted alongside the other fine arts of poetry, painting, sculpture and architecture. Thus, for the first time in history the ground was prepared

> for a clear distinction between the arts and the sciences, a distinction absent from ancient, medieval and Renaissance discussions of such subjects even though the same words were used. In other words, the separation between the arts and the sciences in the modern sense presupposes not only the actual progress of the sciences in the seventeenth century but also the reflection upon the reasons why some other human intellectual activities which we now call the Fine Arts did not or could not participate in the same kind of progress.[72]

For much of the seventeenth and early eighteenth centuries, the doctrine of music did partake of the same kind of progress as that of the natural sciences; but from about 1760 the fine art of music was increasingly regarded as a mystery: at worst, it was thought to be inexplicable; at best, explanation was employed chiefly for pedagogical rather than for theoretical purposes. By the end of the

---

Society, on the Development of the Encyclopaedia, and on the Plans for the Higher Education of the Indians of New England and Virginia...(London, 1932).

[71] For the different arts grouped under the rubric 'polite arts', see Kristeller, 'The modern system' (note 2). In music the useful arts—crafts or trades—included music engraving, printing, selling and the making of musical instruments. The crafts that contributed to the development of modern engineering were eventually distinguished from the trades and classed as mechanical arts, the term 'mechanical' denoting machinery and not the traditional meaning of servile. Mathematics and natural sciences were classed under the rubric 'sciences', along with the various social sciences that originally were part of moral philosophy (e.g., psychology, ethics, politics, economics, anthropology), whereas literature, or *belles letters*, included history in all its branches—even 'fictitious' history.

[72] Kristeller, ibid., p. 194.

eighteenth century, therefore, more and more writers began to speak out against the notion of music as a science. In 1784 Charles Dibdin called into question the mechanistic premisses underlying the practice of musical composition, when he observed that melody and harmony were constituent parts of music but that music with too much harmony was too 'scientific', since harmony was 'a Combination of Sounds upon the most mechanical Principles; and may be shifted, twisted and turned at Pleasure, without the smallest Fancy', whereas 'Melody springs from Genius; and who will be hardy enough to say that Melody is not the governing Requisite in a Song, to which Harmony is only a subordinate Attendant'.[73] In Dibdin's day, however, the prevailing view was that melody derived from harmony and that harmony itself was formed on mechanistic principles.[74] Thus, Samuel Arnold replied that Dibdin might be allowed to speak about harmony

> according to the best of his knowledge, as of a certainty he does to the best of his practice, and that such shiftings, twisting, and turnings in harmony as he has made shift to produce, may indeed be formed without the smallest fancy; but we must be permitted to inform him at the same time, that though harmony is formed on mechanical principles, it requires something more than mechanism to be shifted, twisted, and turned to those effects which move the admiration of a judicious audience, and which we only experience from the greater works of real masters.[75]

The friction between Dibdin and Arnold arose in part from personal antagonism, but it illustrates a conflict that becomes apparent toward the end of the eighteenth century. The conflict is between writers who hold that the operations of musical composition can be scientifically explicated and those who insist they cannot be explicated, since musical composition is the product of genius,

---

[73] C. Dibdin, *A Novel and Interesting Musical Work...intituled the Pupil's Compendium*...(London, c.1784).

[74] For some who held these opposing positions, see Kassler, *The Science of Music* (note 10), vol. 1, pp. 105–108, vol. 2, pp. 823–837, 989–994 *et passim*.

[75] [S. Arnold], *European Magazine* (1785), 7: 183–184, who reviewed Dibdin's work (note 73).

originality and creative imagination.[76] This latter claim brought a strong rebuttal from Augustus F. C. Kollmann, who in 1796 defended his own attempt to deduce rules against the objections of those who argued that 'no rules can be given, that no rules need be given, and that no rules ought to be given for musical composition'.[77] Kollmann, however, is one of the last writers in England to apply rigorous standards of proof and demonstration to his theories of musical composition. Indeed, what began in the seventeenth century as a period of intense activity in the science of music, broadly defined, ends with a period of decline of interest in using the new analytical methods.[78] This decline, which is evident in all fields of musical science—natural-philosophical and moral-philosophical—coincides with the increasing irrationalism of early modern aesthetics, which in turn signals the rise of romanticism and the modern cult of genius that has tended to isolate the composer from his social environment and from the learning of his time.[79] By 1830, therefore, George Hogarth could assert:

---

[76] For an early instance of this conflict, see Kassler, *The Science of Music* (note 10), vol. 1, pp. 381–384.

[77] A. F. C. Kollmann, *An Essay on Musical Harmony*...(London, 1796); see also Kassler, *The Science of Music* (note 10), vol. 1, pp. 649–652.

[78] According to Gillispie, *The Edge of Objectivity* (note 30), pp. 169–170, writers on physical science placed reliance on algebraic resolution of abstract relations of quantity until about the end of the eighteenth century, when they moved away from such a reliance to simple pigeon-holing of phenomenon. See also R. Olson, 'Scottish philosophy and mathematics: 1750–1830', *Journal of the History of Ideas* (1970–71), 32: 29–44, who states that after 1750 analytical mathematics (i.e., algebra) developed on the continent, not in Britain.

[79] The romantics stressed Beethoven's genius but forgot his learning, even though a memoir, published during the composer's lifetime, recorded some aspects of Beethoven's erudition. According to the anonymous author, professional musicians, '(perhaps from the time they must devote to practice, in order to ensure excellence), have not always the reputation of possessing that knowledge in science, which, in the present state of civilization, is inseparable from good education'. Nevertheless, that author could not remember

> having heard more knowledge displayed, than during the few hours he had the good fortune to spend with Beethoven. He has had a regular education, and studied for a considerable time even more abstruse subjects, such as, for instance, the philosophy of Kant.

> Music ... is not a science, nor has it a theory. It is an art—like poetry and painting,—in the exercise of which, we are directed by a number of technical rules, which have no resemblance to the deductions from the theory of a science.[80]

See *A New Biographical Dictionary, of 3000 Contemporary Public Characters, British and Foreign, of all Ranks and Professions* (2d edn., 3 vols. in 6, London, 1825), vol. 1, pp. 125–128.

[80] G. Hogarth, 'Musical literature', *Blackwood's Edinburgh Magazine* (1830) 28: 471–481, who also stated as a 'fact' that no musician of the stature of Beethoven would bother with musical erudition—i.e., musical acoustics or musical antiquities. See also Kassler, *The Science of Music* (note 10), vol. 1, pp. 523–524.

8. MUSIC AS MATTER IN MOTION

... all differences of forms of matter, depend on motion.
R. Descartes, *Principia philosophiae* (1644, II.xxii)

Introduction

This chapter examines the work of a little-known music theorist, William Jones of Nayland (1726–1800). I shall describe what there is relating to musical form in Jones's main writings on music. Once these data are given, I shall examine the rule of construction implicit in his theory.[1] A rule of construction is an enunciation of the supposed and evident conditions imposed upon music-theoretic knowledge by the fundamental ideas which that knowledge involves and which contain in themselves the elements of those 'truths' that Jones discovers and enunciates. When understood in their logical connection, Jones's fundamental ideas enable us to characterise explicitly his theory of music. While this characterisation will derive

[1] By 'rule of construction' I mean something similar to the term 'construct', coined by Henry Margenau to mean more than is implied by a concept or abstraction. According to him, concept formation—the union of particulars into universals—occurs at all levels of a science; but it is 'an elementary form of construction, and only an elementary form'. Construction is an addition to concept formation, for it endows 'the product with suitable properties of its own: it is a creative as well as a synthetic act'.

> The passage to orderly knowledge involves the positing of constructs, which are the rational elements to which datal experience is made to correspond ... To indicate that they are not mere gleanings from the field of sensory perception, that they come into their own through what are felt to be creative processes in our own experience rather than through passive contemplation; to emphasize their rational pliability and yet to distinguish them from shadowy concepts, the writer has ... called them *constructs*... .

See H. Margenau, *The Nature of Physical Reality* (New York, 1950), pp. 68–73.

from the theory itself, it will refer also to some of Jones's other writings and ultimately to the broader conceptual background from which his theory arises and from which it draws its validity. This background will provide the means by which Jones's theory might be tested—by which we might not only understand what is there but also why some things are there and others are not.

William Jones was baptised on 20 July 1726 in the parish church of Lowick, Northamptonshire.[2] On 17 September 1740 he was admitted to the Charterhouse, London, where he received a classical education and developed interests in natural philosophy and music. He may have studied music theory with Johann Christoph Pepusch, organist of the Charterhouse Chapel; but he certainly studied privately with the musician, James Oswald, who had settled in London in 1741.[3] Jones also formed an important friendship with Charles Jenkinson, afterwards 1st Earl of Liverpool, who was a 'chum' at Charterhouse and, later, at University College, Oxford, where Jones matriculated on 9 July 1745 and where on 19 July he received a Charterhouse Exhibition (a sum of money to aid poor boys during their university careers). During his Oxford years Jones learned about Jean-Philippe Rameau's doctrine of fundamental bass; and he also became acquainted with the writings of John Hutchinson, which he and his friend, George Horne, afterwards Bishop of Norwich, studied with assiduity. On 8 April 1749 he was awarded the degree of B.A.; in the same year he was ordained a deacon and afterwards, in 1751, a priest.

On leaving the University in 1750, Jones was presented to the curacy of Finedon, Northamptonshire. Although he held this post until 1760, after his marriage in 1754, he removed to Wadenhoe, where he was curate to his brother-in-law, Brook Bridges. During this period he conducted experiments in natural philosophy, which were

---

[2] See W. Jones, *The Theological, Philosophical and Miscellaneous Works* ed., with a life, by W. Stevens (12 vols. in 6, London, 1801); see also J. C. Kassler, *British Writings on Music, 1760–1830...*(2 vols., Columbia University doctoral dissertation, 1971), *The Science of Music in Britain, 1714–1830: A Catalogue of Writings, Lectures and Inventions* (2 vols., New York, 1979), vol. 1, pp. 589–607, 'William Jones', *The New Grove Dictionary of Music and Musicians* ed. by S. Sadie (20 vols., London, 1980), vol. 9, p. 705.

[3] Jones, *The... Works* (note 2), vol. 10, p. 72.

supported for three years at 300 pounds per annum by the subscription of friends. The resulting publication, *An Essay on the First Principles of Natural Philosophy* (Oxford, 1762), was admired by John Stuart, 3d Earl of Bute, who authorised Jones to charge to his account any expenses incurred in the purchase of mathematical instruments and who made his private library accessible to Jones. Reception of the *Essay* in the periodical literature, especially the *Monthly Review*, was less favourable, because Jones had attacked doctrines favoured by the Newtonians.

In 1764 Jones was presented to the vicarage of Brotherton and in 1765, to the rectory of Pluckley, both in Kent. In 1775 he was elected a Fellow of the Royal Society, and in August of the following year he embarked for France. The French language presented no difficulties, because Jones had been reading French 'for many years'.[4] Except for the journey to and from Paris, most of his time was spent in the city, visiting libraries, playing the organ in Notre Dame, browsing in natural history museums and taking in other sights of interest. Not long after his return to England, Jones removed to Nayland, Suffolk, where from 1777 until his death he made his home.

In 1782 Sidney Sussex College, Cambridge awarded Jones an M.A., for he had 'lost the gift of a degree' from Oxford because of his controversial *Essay*.[5] The Royal Society appointed him preacher of the Fairchild sermons on natural history, which he delivered annually between 1784 and 1787 at the church of St. Leonard Shoreditch, London. About this time he seems to have been a member of a club that met in Colchester to discuss literary subjects, the leading members of which were Samuel Parr and Thomas Twining, the latter of whom was much interested in music.[6] Ever solicitous for the well-being of his parishioners, Jones studied physic, composed music for

---

[4] [W. Jones], *Observations in a Journey to Paris by Way of Flanders*...(2 vols., London, 1777), vol. 1, p. 22.

[5] See his letters to James Boswell (4 January 1794) and the Hon. George Kenyon (2 September 1794), *Historical Manuscripts Commission, 14th Report, Appendix, Part IV. The Manuscripts of Lord Kenyon*...(London, 1894), pp. 538–540, 548–549.

[6] For Twining's contributions to music, public and private, see R. Lonsdale, *Dr. Charles Burney: A Literary Biography* (Oxford, 1965).

the benefit of the church at Nayland (and to try out his music-theoretic principles) and took pupils. When he was no longer able to carry on as a tutor, he was presented to the sinecure rectory of Hollingbourne, Kent, in 1798. His will is dated 13 May 1799; it was proved 22 February 1800.

Jones was a man of strong and unwavering convictions, so that some of his writings on religion and politics convey a tone of intolerance.[7] But he may have been more personable in private life, since Twining, in a letter to Charles Burney, wrote that Jones was 'a very sensible, agreeable man; but tinctured with some prejudices very unphilosophical for a Philosopher'.[8] And Cecilia Maria Henslowe (*née* Barthelémon) wrote that when her father, the musician, François-Hippolyte Barthélemon, converted to a dissenting religion—Swedenborgianism, although the family was dismayed, her father and his particular friend, '(that late pillar of the church) Trinity Jones of Nayland', passed many agreeable hours together in religious conversations.[9]

The nickname 'Trinity Jones' referred not only to Jones's orthodox belief in the doctrine of the Trinity but also to his use of tripartition as proof of the Trinity. For example, he posited three principles of motion in the universe (fire, light, air) and in man (heart, lungs, nerves); and three principles of beauty in painting (red, yellow, blue), in architecture (length, breadth, height) and in music (keynote, third and fifth). But man also has three principles, because, as a musical instrument of God's forming, he has music in his voice, in his ear and in his whole frame.[10] Hence, for Jones, the 'Physical

---

[7] See W. Jones, *The...Works* (note 2), for evidence that Jones was a zealous opponent of dissenters, seeing in them a challenge to the established church and the established government. But his ire seems to have been directed especially against Joseph Priestley, John Locke and F. M. A. de Voltaire.

[8] See R. Lonsdale, 'Dr. Burney and the *Monthly Review*', *The Review of English Studies* (1963) 14: 346–358, p. 354.

[9] [C. M. Henslowe], 'Memoir of the late F. H. Barthélemon', *Selections from the Oratorio of Jefte in Masfa, composed by the late F. H. Barthélemon* (London, c.1827), pp. 5–10.

[10] W. Jones, *The Nature and Excellence of Music. A Sermon preached at the opening of a New Organ, in the Parish Church of Nayland in Suffolk*...(London, 1787), reprinted in *The...Works* (note 2), vol. 6, pp. 110–127.

Trinity', as an absolute fact in music, must be evident to every beginner in the science'.[11]

William Jones's Representation of Music

Jones wrote two treatises on music. The first, 'On the Philosophy of Musical Sounds', is one of the 'discourses' in Jones's *Physiological Disquisitions,* a book on the physical universe. Hence, this treatise falls within the domain of natural philosophy.[12] Although its principal subject is the physics of music, Jones digresses to consider an aspect unrelated to physics, namely, 'method or style in music, without which the most laboured composition is a rhapsody of nonsense'.[13] Despite this digression, the treatise presents the chief of Jones's 'speculations' on the subject and, hence, is of little use to a practitioner. By contrast, the second work, *A Treatise on the Art of Music*, presents a practical theory, for Jones's purpose is

> to lead the Musical Student by the shortest and easiest way through the principles of Consonance, Modulation, Harmony, and Musical Air, to an understanding of what others have done, and are doing in the Art; and to lay down such Rules as will enable him to acquire a learned Taste, and to judge of the comparative merit and excellence of their productions.[14]

---

[11] Jones, *The...Works* (note 2), vol. 10, p. 98.

[12] W. Jones, *Physiological Disquisitions; or, Discourses on the Natural Philosophy of the Elements...*(London, 1781), pp. 293–358; reprinted in Jones, *The...Works* (note 2), vol. 10, pp. 1–100. The eight sections in the treatise are: Of sound in general; Of musical sounds; Of the scale of music; Of the monochord; On sympathetic sounds and harmonics; On the Eolian harp; On style and method in music; Some reflexions on the uses and application of music.

[13] Ibid., p. 346.

[14] [W. Jones], *A Treatise on the Art of Music; in which the Elements of Harmony and Air are practically considered...*(Colchester, 1784), pp. 57–58. A second edition was published posthumously in 1827 with the author's name.

In this second treatise the subject is the intellectual world, namely, musical ideas communicated by composers via musical pieces. Hence, the treatise falls within the domain of moral philosophy.

In both treatises Jones claims to deduce his principles from facts—in the one, vibrating strings and in the other, musical compositions.[15] And in both treatises Jones claims to identify genuine knowledge with the fruits of scientific method, the rules for the conduct of which were to start from observation—'intentional experience'. From this starting point, a writer was supposed to reason and draw conclusions directly from phenomena and then, applying those laws, account for other things.[16] When applied to his second treatise, this method served Jones's object of setting forth the principles and rules for understanding, critically, the practice of musical composition.[17]

To achieve this object, Jones begins with music's foundation or 'substratum', namely, the 'concordance of musical sounds, and their combinations', that is, chords and their progressions. He defines progressions as 'a series or chain of Chords, connected together, and depending on one another till they come to some kind of Close'. And he describes and illustrates the progressions in their diatonic and chromatic species so as to show the several combinations and permutations of chords that may constitute a variety of different harmonic periods. In this way Jones endeavours 'to exhibit the most considerable parts of what may be called the *Materia Musica*, which experience and fancy are to compound and apply in many different ways':[18]

---

[15] In *Physiological Disquisitions* (note 12), p. xix, Jones notes that his work 'is properly *Physiological*', because its 'demonstrations are from facts'. In *A Treatise* (note 13), p. v, he states that his 'principles are rather extracted from Musical Compositions than from Musical Treatises'.

[16] See, e.g., *The Modern Encyclopaedia* ed. by A. D. Burrowes (2d edn., London, 1822) *s.v.* 'Newtonian philosophy'.

[17] Jones, *A Treatise* (note 14), p. x, cautions that his moral-philosophical treatise 'cannot be skimmed over like a Novel, but demands the thought and attention of a philosophical or mathematical treatise, in those who wish to profit by it'.

[18] Ibid., p. 40.

> By more changes and inversions, and by looking farther into the works of some of the best Authors, I might have multiplied these examples; but when the Learner has studied what I have here given, he will certainly be able to do this and much more for himself. Out of these periods, he will construct others by compounding them together at pleasure, when use has made it easy to apply the rules of Inversion and Modulation: and when fancy is furnished with some variety of Air and Measure, the Learner becomes an *Extempore*-Performer, who will find his stores inexhaustible, if he has laid in a proper foundation of Harmony.[19]

Although the 'Art of *Composition* includes *Invention* and *Disposition*'; 'the latter only can be taught ... [and] supposes a previous knowledge of Topics, that is, of the Common-places, Predicaments, and Figures, of Music'.[20] By topics, Jones denotes the rules for utilising chords (commonplaces), their disposition (predicaments) and their progression or harmonic periods (figures). These rules relate to music's substratum, upon which the 'superstructure' of air and expression is built. Although that superstructure is the product of untaught nature—'fancy or Imagination', it still may be placed under certain restrictions, since 'there is no Science, nor any one part of a Science, in which Nature will not derive much assistance from Art and Experience'.[21] Method, therefore, is necessary to all parts of a musical 'discourse', whether those parts stem from invention (nature) or disposition (skill). Hence, Jones points out that if his work

> has any merit in which it differs from all others of the kind, it will be found in the eighth Chapter on the Analysis of Air and the Conduct of Subject, where I have attempted to write a new Syntax of Melody and Musical Rhetoric; not new in itself, because it is exemplified from the Works of Masters who could not have followed the Rules of it unless

---

[19] Ibid.

[20] Ibid.

[21] Ibid., p. 41.

they had known them; but by new I mean such as has not to my knowledge been professedly explained.[22]

In providing the 'new Syntax', Jones treats air as 'the manner in which Harmony is expressed', and melody as 'Harmony in succession', 'a solution of notes from their Harmonic Consonance'.[23] But air and melody are not synonymous terms, since, in addition to melody, air requires measure; and this latter constituent respects both the number and duration of notes in a bar and the length of a musical 'versicle or clause' or a series of correspondent 'clauses in a Strain'. When melody and measure are compounded, the term 'air' denotes subject, for Jones points out that pieces of music 'written upon no subject can have no sense'. Hence, the subject of a piece 'must be proposed in some certain measure; and, in the clauses and periods that follow, nothing should be introduced that cannot be reconciled either with the primary, or with some secondary subject, which it is lawful to introduce after a rest or cadence'.[24] The principal and the subordinate subjects give the piece its 'formality', which a skilful composer conceals by varying the periods 'at discretion, and ... interposing rests and artificial interruptions'.[25]

Just as music's foundation—harmony—is disposed according to certain rules, so too is music's superstructure—the subject. In the case of harmony, the rules assist a composer with harmonic progressions, including transitions to other keys. But in the case of the subject, the rules assist in framing the subject with proper emphases and proper stops so that a composer may express 'the Idea or the passion which he proposes to excite by any particular passage or movement'.[26] Framing consists of techniques equivalent to figures of rhetoric—for example, climax, antithesis, periphrasis, parenthesis,

---

[22] Ibid., p. ix; see also p. 41, where Jones writes: 'the attempt to reconcile Air with Reason, will appear like that of giving Laws to the Wind, *which bloweth where it listeth*; but when the matter is properly considered, it will appear in a different light'.

[23] Ibid., p. 41.

[24] Ibid., p. 46.

[25] Ibid.

[26] Ibid., p. 41.

aposiopesis, which Jones singles out as corresponding to the musical figures of augmentation, diminution, inversion, reversion, iteration, division and imitation.[27] This last figure is not imitation of one passage in another passage, 'or the imitation of Music in Music, but the imitation of nature in musical sounds'.

The various techniques of expression make up the principal rules for the conduct of subject, which thus considered may seem 'very formal and simple: but every thing is formal in a certain degree which is methodical and all complicated things become simple when they are decomposed'.[28] Jones illustrates the rules by the varieties in which a subject may be ordered, that is to say, the places in which a 'stop' may be made. In common time, for example,

> it is very usual to make a stop, equivalent to a Comma, after three fourths of the first Bar; another stop, equivalent to a Colon or Semicolon, after three fourths of the second Bar: the first clause containing the first or principal subject, the next a second or subordinate Subject. Sometimes the first comma is found at three fourths of the second Bar: and another stop at the correspondent part of the fourth Bar; or, in pieces which allow a latitude, the Composer omits the second stop, and continues the air at his pleasure, to avoid formality; yet under such restriction, that the number of Bars to which a musical period is extended, may be some Multiple of the leading, or of some secondary subject.[29]

The 'Clauses of Air, and the successions of melodies' constitute 'antecedents' and 'consequents', and these have a mutual relation and agreement in several respects. In some cases, the consequent may

---

[27] Ibid., p. 43. Climax, or marching figure, is the mounting by degrees through words or sentences of increasing weight and in parallel construction. Antithesis is the conjoining of contrasting ideas. Periphrasis is a word, phrase or sentence inserted as an aside in a sentence complete in itself. Parenthesis is a word, phrase or sentence inserted as an aside in a sentence complete in itself. Aposiopesis is silence or interruption, which is effected either by stopping suddenly in mid-course, leaving a statement unfinished; or by leaving an idea unexpressed but clearly understood. See R. A. Lanham, *A Handlist of Rhetorical Terms: A Guide for Students of English Literature* (Berkeley, 1969).

[28] Jones, *A Treatise* (note 14), p. 48.

[29] Ibid., pp. 46–47.

agree in harmony with the antecedent, thereby allowing its combination with said antecedent. In other cases, the consequent may be a repetition of the antecedent, or it may be a repetition of the antecedent in the next related key and in another part. The consequent may further agree with the antecedent in its air, its measure, as well as in its similar course of modulation in another key.

But Jones points out that sameness and repetition are 'disgusting in every kind of composition' and are avoided in music 'by certain Methods of diversifying the original Air without departing from the sense of it'.[30] He identifies four means of doing this: (1) the air may be broken into parts, 'each of which may be taken up and pursued in its turn at the fancy of the Composer'; (2) two different subjects may be proposed in an upper and under part and interchanged; (3) one may retain the notes of the melody and the order of them but change the value of the time either by augmentation or diminution; and (4) the notes of the air may be transferred from a major to a minor key or vice versa.

From the above analysis of Jones's argument, it is possible to identify his three fundamental ideas or hypotheses. First, he considers music as a particulate mass, divisible into various classes of particles—notes, intervals, chords, harmonic periods and subjects. Second, he regards these classes as material to be used—for example, he considers the subject for what it can do, specifying two functions: the production of uniformity and variety. Both functions arise from Jones's ideas of repetition, including reiteration or immediate repetition and recurrence or repetition at some later point in time. Here, then, we have Jones's first principle of music, which is clearly stipulated in his other, natural-philosophical treatise on music: 'All the beauty with which we are justly charmed in the works of art, consists in variety with uniformity'.[31]

Third, from this principle Jones deduces a general rule for the conduct of subject: 'Every musical subject must ... be under the restriction of isochronism and commensurability'.[32] Isochronism

[30] Ibid., p. 50.

[31] Jones, *Physiological Disquisitions* (note 12), p. 328 [sic = 348].

[32] Ibid., p. 347.

pertains to the equality of subjects as to time, metrical length, duration, intervals of occurrence and the like, whereas commensurability has to do with the reducibility of subjects to a common measure. Whether these restrictions are taken as cause (means) or effect (ends) in the production of uniformity and variety is purely a question of the order in which Jones takes them relative to one another. By hypothesis, he attributes to the subject general immutable properties, and these properties are causal properties. Representation and understanding of the general principles of causal relation commit Jones to consider a subject in relation to other things and never in and for itself. Since subjects are instances of a class in music, he regards them as bare units and equivalent, sharing the common property of causality.

Hence, Jones's rule of construction is this: *a musical piece may be represented symbolically as a unity produced by the repetition of subjects*. This rule commits Jones to ignore facets of subjects that are non-causal, so that the only differences accounted for within a musical piece are the change in the arrangement of subjects or the order in which they follow one another. The patterns of repetition that can be constructed upon the basis of this rule stem from pure mathematics, as shown in *Figure 8.1*. Thus, any piece of music may be known by the original or primitive subjects. By knowing these, the entire musical piece is known. This is so because Jones represents a piece of music as the sum of subjects. Since all the subjects, once they make their appearance, are represented as being repeated, the only facet that will account for this piece of music as distinct from that piece is the order and arrangement in which the known beginnings are conducted.

In his review of Jones's moral-philosophical treatise, Charles Burney took particular issue with this rule of construction, pointing out that Jones confined the production of a musical piece 'wholly to the repetition of the same traits of melody in different keys throughout the piece, from the beginning to the end; which he

UNIFORMITY Imitation of Subject*	VARIETY Variation of Subject	
	Music	*Mathematics*
1. Repetition on the same (or on a different) note of the scale	1. Change of key and mode	1. Change by transposition
2. Repetition in notes of the same (or of different) lengths	2. Change of measure	2. Change in magnitude by augmentation or diminution
3. Repetition in the same (or in a different) motion	3. Change of movement	3. Change in shape by inversion or 'reversion'
4. Repetition in the same (or in a different) form	4. Change in the divisions of the harmony and air	4. Change in form by division

* Imitation includes the ideas of same and similar, that is, strict and free imitation.

Figure 8.1. Patterns of repetition

recommends under the notion of adhering to a theme or subject'.[33] But the problem in Jones's theory is more complex than Burney suggests, for a musical piece, considered as an arrangement of equivalences in musical space and time, contains nothing to account for the represented change. The arrangement of subjects does not account for the change, for the space in which they are arranged makes no qualitative difference of any kind to them.[34] The subjects themselves do not explain the change, since they are regarded as bare units and equivalent. Therefore, the source of change in a musical piece must be looked for eventually outside the complex. However much Jones includes in his system of musical symbols, such a representation refers, at some point, beyond itself to an external cause.

In Jones's theory the point at which the purely musical representation breaks down is in the relationship of genius to the rules. Whether Jones defines genius as fancy, feeling, imagination, taste or invention, this attribute is a property of the soul, a God-given gift.[35] While such a gift may be improved by experience, it cannot be accounted for in an explanation of music. The form of Jones's thought compels him to refer to something beyond the system of musical symbols; and the process moves from the hierarchy concerning the material of music to its purpose—to the composer and his intention and, ultimately, to God.

THE CONCEPTUAL BACKGROUND OF JONES'S THEORY

Under the pattern of thought I have been describing, change is represented as being completely determined. But to utilise the word 'determinism' as descriptive of the kind of thought exhibited in the

---

33 [C. Burney], *Monthly Review* (1786) 75: 105–112, 174–181, p.177.

34 E.g., Jones does not prescribe an ending time for a piece of music, nor does he prescribe the maximum number of subjects (or other elements of music) of which a piece of music should consist. Thus, musical space and time do not appear to be axiomatic.

35 See, e.g., Jones, *The…Works* (note 2), vol. 6, p. 111, who asserts that music needs no other 'recommendation to our attention as an important subject, when it shall be understood … that it derives its origin from God himself'.

writings of Jones would be misleading, for such a word suggests the disability to function without an agent or the necessity of being set and kept in motion from without. That is, determinism would refer to the characteristics of instrument or machine and to the inanimation proper to a machine. In practice it would be very difficult to test the character of Jones's theory by this criterion.[36] If we take the word 'deterministic' to mean 'with the aid of the concepts of mechanics', then the term 'mechanistic' might be more accurate in characterising the nature of Jones's theory. Mechanics—not mechanisms—in this case would be taken in the sense of the doctrine of the motions of material bodies.[37]

That doctrine included mathematical treatment not only in the sense that it made use of mathematical terms and method for abbreviating arguments but also in the more stringent sense that its basic concepts were mathematical concepts—that mechanics itself was a mathematics.[38] In Jones's day, the name most frequently associated with mechanics was that of Isaac Newton,[39] although the

---

[36] Musicologists tend to apply the epithets 'rational' or 'empirical' to eighteenth-century theories of music. See, e.g., P. H. Lang, *Music in Western Civilization* (New York, 1941), p. 441, and L. F. Chenette, *Music Theory in the British Isles during the Enlightenment* (Ohio State University doctoral dissertation, 1967). A rational procedure denotes that which utilises hypotheses from which facts can be deduced, whereas an empirical procedure is that which compares the range of data of different kinds, where analysis is with reference to several suitable examples. Deductive logic has an important place in the former method and inductive logic in the latter. Jones claims to utilise the empirical method but in fact employs the rational. To characterise his theory by one or the other epithet would neither provide us with insight into the content of his theory nor provide us with a means of testing that theory.

[37] See, e.g., Margenau, *The Nature of Physical Reality* (note 1), p. 178: Mechanics is 'a discipline that deals with all experience that can be represented and understood with the use of three kinds of *system*: (1) mass points (more loosely called particles) or sets of mass points; (2) rigid bodies; (3) deformable, continuous material media ... [These three] systems are often considered in motion, but the conditions under which they are at rest are just as important as is a description of their motion'.

[38] See E. J. Dijksterhuis, *The Mechanization of the World Picture* (Oxford, 1969), pp. 463–491.

[39] Newtonian mechanics focused upon the particle, a general concept that included the atom, emphasising mass, position and velocity as essential

formulation of the mechanical philosophy of matter varied in structure and detail during the eighteenth century.[40] Jones, however, belongs to the only school of British anti-Newtonians: the Hutchinsonian school named after its founder, John Hutchinson, whose writings Jones had studied at Oxford.[41] Although Hutchinson claimed to derive the physical world from his reading of Scripture, particularly the Old Testament in its original Hebrew, this reading had roots in French literature, especially the writings of Descartes and his followers.

While Hutchinsonian ideas are most clearly apparent in Jones's first book, the controversial *Essay* of 1762, they also pervade his second book, the *Physiological Disquisitions*. Dedicated to his friend, Jenkinson, this work includes, besides the natural-philosophical treatise on music, other 'discourses' on matter, the nature and causes of motion, the nature and properties of air, fossil bodies, physical geography, and the appearances, causes and prognostic signs of the weather. Jones therefore understands the word 'physiology' broadly, as an investigation of the causes of physical phenomena in order to construct a theory of nature. In the course of his investigation, he sought to reduce natural phenomena to one simple and universal law. This law is the natural agency of the elements—matter in motion. Jones understands the 'first particles or principles of which matter is composed' as impenetrable '*atoms*, those very small invisible elements of bodies which cannot be *cut* or *divided*'.[42] But he also understands atoms in a grosser sense as well, 'as constituent parts of

attributes of the particle. But the two key notions of Newtonian mechanics were (1) a vacuum, i.e., unfilled space, and (2) the binding principle of gravity.

[40] For some of the varieties, see R. E. Schofield, *Mechanism and Materialism: British Natural Philosophy in an Age of Reason* (Princeton, 1970).

[41] For the principles of Hutchinsonian mechanics, see ibid., pp. 122–128, 132–133, 164. For a short statement of Jones's Hutchinsonian principles, see W. Jones, *Memoirs of the Life...of the Right Reverend George Horne*...(London, 1795); see also the new preface to ibid. (2d edn., London, 1799). For Jones's Hutchinsonian milieu, see W. K. Lowther Clarke, *Eighteenth Century Piety* (London, 1945), pp. 101–117. For the wider background, see C. B. Wilde, 'Hutchinsonianism, natural philosophy and religious controversy in eighteenth century Britain', *History of Science* (1980), 18: 1–24.

[42] Jones, *Physiological Disquisitions* (note 12), p. 1.

bodies ... distinguished by their figures'.[43] The figure of the whole 'arises from the figure of the parts', and those parts may be divided from each other 'without any limits which we are able to determine':[44]

> How far soever matter may be divisible in its own nature, we can conceive no idea of its composition, but as it consists of units ... We therefore find ourselves under the necessity of considering all matter as made up of atoms or units; and it is most reasonable to suppose, that all bodies, which are the same kind, agree in the configuration of their smallest parts, and consist of elements rendered permanent and unchangeable by the express design of the Creator... .[45]

Change—variety—is due not to the atoms themselves but to their configurations, for when 'the atoms of matter are assembled into masses ... many new properties arise from their different associations'.[46] Hence, Jones traced the phenomena of nature back to their principles in the laws of motion, by means of which the universe is considered as

> a great machine, subject to a variety of motions, some of which are constant, and others variable and temporary. All the elements are in motion ... [and if] we inquire after the first and original cause of all this motion, that cause can be no other than the power of God... .
>
> But when we consider God as the cause of motion, it is not to be understood that he acts as the *immediate* cause in all the operations of nature.... Observation teaches us, that there is in the world a chain of causes, each depending on the other, and all subordinate to the power of the Creator. To follow this chain of causes, either by descending from the greater to the less, or by ascending from the less to the greater, is the proper business of philosophy ... so far as the system of nature is intelligible by us, it must be considered as other machines.[47]

---

[43] Ibid., p. 2.

[44] Ibid., p. 3.

[45] Ibid., p. 5.

[46] Ibid., p. 9.

[47] Ibid., pp. 27–28.

This construction of matter in motion is indebted to the mechanical philosophy of the Cartesians who equated matter with extension and 'denying the existence of the void, filled space with variously sized and shaped particles resulting from the infinite divisibility of dimension. To these they added a fixed and finite quantity of motion which was distributed and redistributed among the particles jostling against one another in the universal plenum'.[48] The identity of matter and space in Jones's theory, as in Cartesian mechanical philosophy, stems from a belief in the infinite divisibility of matter, for a space not containing any matter would be a contradiction and, consequently, impossible. While preserving their shape, the particles of matter move relative to each other; hence, the differentiating principle for Jones, as for the Cartesians, stems from relative motion and relative rest which cause the parts of space to form a coherent whole and to differ from other parts.

In his writings on matter in motion, Jones did not utilise mathematics *per se*, for his arguments are expressed in the language of everyday speech. Even so, his thought may be described as mathematical, though it does not necessarily employ pure mathematics as its instrument. What characterises Jones's thought, however, is his description of nature as cause. In his natural-philosophical treatise on music, the word 'nature' refers to the physical cause of sound; in his moral-philosophical treatise, where sound is considered as an effect already achieved, nature refers to the intellectual cause of music. In both treatises Jones utilised a description that is aided by the mathematical concepts of Cartesian mechanics, as shown in *Figure 8.2*; and the reasoning based upon these concepts may be regarded as mechanistic.

Fundamental to mechanistic thought is the analysis of an object into a set of units and the arrangement of these units in an order. The abstractions employed in an analysis of this sort force upon Jones a mechanistic interpretation of action. Any interpretation based upon this type of symbolic representation must represent music as a complex in which all action is completely determined in accordance with causal laws. These laws, and the mechanistic reasoning employed in exhibiting them, stem from two sciences that

---

48 Schofield, *Mechanism and Materialism* (note 40), p. 7.

MECHANICS	PHYSICAL WORLD	INTELLECTUAL WORLD
Substructure (at rest)	Matter	Harmony
First principles	Atoms	Notes (rests)
Grosser particles	Various configurations of atoms	Intervals, chords, periods
Superstructure (in motion)	Figure	Subject
Rules for conduct	Principles of uniformity and variety (see *Figure 8.1*)	Principles of uniformity and variety (see *Figure 8.1*)
Rules for differentiation	Relative motion and relative rest (i.e., arrangement of, or order in, which figures follow one another)	Relative motion and relative rest (i.e., arrangement of, or order in, which subjects follow one another)
Secondary Cause	Physical nature	Human nature
Primary Cause	God	God

Figure 8.2. Jones's mechanical philosophy

implemented the two 'beliefs' of Cartesian mechanics: geometry assisted in the belief that entities (systems) are divisible into parts, and kinematics assisted in the belief that these parts are localisable in space and time. The axioms of geometry express the necessary conditions that result from the idea of space, while the axioms of kinematics express the necessary conditions that flow from the ideas of motion.[49]

In mechanics the relationship of these two sciences arose from the properties or 'affections' of both. These properties, which constituted the mechanical affections of matter and form, were of two kinds: (1) primary affections 'which arise either from the idea of matter, as quantity and figure; or from that of form, as quality and power; or both together, as motion, place and time'; and (2) derivative affections which arise from some of the primary ones, as for example, 'from quantity, as divisibility, continuity, contiguity, finity [sic], impenetrability; from figure, as regularity and irregularity; from quality, as strength', etc.[50] Affections of the mind—passions— were treated as derivative.

If mechanical properties arose either from the idea of matter or from that of form, we might ask: what is matter and what, form? Matter denoted any physical substance not definitely particularised. Form was the manner of being peculiar to each body and which distinguished it from every other. Matter was the common basis, the substratum; form was that which specified and distinguished, the superstructure. In Jones's moral-philosophical treatise on music, matter is harmony—chords and their progressions, whereas form is air and expression—that which shapes harmony into particular pieces of music. Such a distinction leads to five different kinds of musical form.

(1) *Natural Form*; *Artificial Form*. Natural form is that which is inherent in nature without anything contributed thereto on the part of humans. Artificial form is that which arises from human industry. In Jones's moral-philosophical treatise on music, natural form is sound

---

[49] With Newton, dynamics enters the equation, the axioms of which express the necessary conditions that flow from the ideas of force and motion.

[50] *The Cyclopaedia; or, Universal Dictionary of Arts, Sciences, and Literature*...ed. by A. Rees (44 vols., London, 1802–1820), *s.v.* 'Affections'.

already achieved; and artificial form is the practical disposition of this matter by means of human skill.

(2) *Simple Form*; *Compound Form*. Under this notion of form, matter is distinguished into parts, as, for example, in Jones's theory, where simple form denotes a single element or particle within a class—a note, a chord, an harmonic period; and compound form is the mixing or compounding of these elements. Or, if one considers a quality of matter, simple form may refer to the duration of a note; compound form, to durations in combination.

(3) *Primary Form*; *Secondary Form*. Primary form consists in figure, extension and organisation, whereas secondary form is derived from primary. The difference between this kind of form and the second kind described above is in the addition of a new construct, organisation, whereby the parts of matter are placed in some kind of order. For example, when used to denote the external appearance of a piece of music by describing the disposition of its parts as to length, breadth and thickness, primary form is synonymous with the term 'figure'. Secondary form refers to a figure's regularity, as in the case of Jones's restriction of subjects to isochronism and commensurability. A concatenation of figures, or subjects, produces extension, which, as a temporal consideration, may include such ideas as frequency and duration. These ideas refer to number, namely, to repetition. The idea of repetition notices figure in its reiteration and its recurrence. But figure may be considered spatially as well, and this consideration may include such ideas as magnitude, or the augmentation and diminution of the figure in repetition.

(4) *Form synonymous with Class*. Form used in this manner refers to all the differently sized units accounted for by a theorist in the construction of a theory or analysis of a piece of music. Taking, for example, three arbitrarily chosen units—small-scale, middle-scale and large-scale structures—within these units some eighteenth-century theorists will account for more, some for fewer classes of particles. Jones, for example, does not treat large-scale structures. The largest configuration of which he treats is a middle-scale structure: the subject. Of course, form as class may refer to other hierarchies in

a theory than structural ones; but I have not included an examination of these facets of Jones's theory.[51]

(5) *Form as a Manner of Doing*; *Form as a Manner of Being*. Form utilised in the first sense refers to the doing of something according to rules. It is an activity, where form refers to the uniting or joining of several different things to make one whole or compound. Form taken in this way denotes the same thing as synthesis—disposing or conducting the various parts of a piece of music according to some regular order. But form as a manner of being denotes a substantive—the compound or thing itself—that is, the piece of music, where the whole is produced by the order and fitness of all its parts.

Conclusion

Although, compositionally, the late eighteenth century seems to have chosen the sonata as its ideal musical form, music-theoretically, it tended to idealise the form and technique of fugue. Jones himself remarks that 'the Fugue ... is the most learned species of Music, and strikes the memory and judgment, with its proportion and symmetry, while it amuses the fancy with Air, and fills the Ear with harmony'.[52] Yet, contrary to Jones's own claims,[53] his description of music does not in fact arise from the empirical data of an existing piece or collection of pieces of music. Rather, his description arises from a particular kind of concrete situation, namely, an interest in using things. The conception of things is the conception of these things as material; therefore, Jones regards the classes of musical elements as analogous to material bodies—*materia musica*. The nature of his representation allows him to account for certain general properties of music but not for particular instances. Hence, all pieces of music—regardless of genre—are analysed and accounted for in the same way.

---

[51] For the structural level relating to the composer's intention (purpose), see Kassler, *British Writings on Music* (note 2), vol. 1, part 2 and vol. 2, pp. 114–121.

[52] Jones, *A Treatise* (note 14), p. 49.

[53] Ibid., p. 94; see also above (note 15).

Jones identifies five properties as constituents of all pieces of music: (1) equivalences, (2) symmetry, (3) mechanism, (4) form relative to number and arrangement of properties and (5) extension. He does not account for such things as (1) opposites, (2) asymmetries, (3) organism, (4) form as a synthesis of opposites and (5) development. Representation of a piece of music as a unity of differences or opposites, proceeding dialectically, is an organic apprehension, not a mechanistic one. A mechanistic interpretation of the type utilised by Jones will account for differences merely as changes in the number or in the arrangement of musical equivalences in space or in time. Symmetry is paramount under such an interpretation: since a piece of music is represented as the sum of equivalences, asymmetry (irregularity) is regarded as a license or altogether disregarded.

All general theories, of course, will omit certain facets of the data to which they refer. The question to ask about such theories is: how well do they represent and account for those data? Jones's data are compositions ranging from the time of Josquin to the time of Haydn, but he focuses chiefly on the 'classical' compositions of Corelli and Handel, because these fit more nearly his representation of music than does the 'modern' compositions.[54] Indeed, Jones's mechanistic reasoning is so far removed from the music of Boccherini and Haydn that pure intuition made the distance evident. For him, these composers 'are sometimes so desultory and unaccountable in their way of treating a Subject, that they may be reckoned among the wild warblers of the wood'.[55]

---

[54] Jones, ibid., pp. iv and 43, admits his bias.

[55] Ibid., pp. 49–50.

9. MUSIC AS UNFOLDING LIFE

In every living germ is a creative idea
which develops and exhibits itself through organisation.
C. Bernard, *Introduction...de la médecine expérimentale* (1865)

INTRODUCTION

It is well known to historians of science that two fundamental philosophies have emerged in history of which now the one—mechanism—now the other—organicism—has had the upper hand in explanations of phenomena. It is less well known, however, that both philosophies are to be found in writings that purport to explain the structure of music. During most of the seventeenth and eighteenth centuries, musical structure was interpreted mechanistically; but the first intimations that music unfolds purposively like a living organism appear as early as 1770 in statements such as the following: 'The main forms of music ... are products neither of chance nor convention; they derive from the laws of nature, in other words from *our* organic structure which makes them necessary, unchangeable and universal'.[1] Some adumbrations of both mechanist and organicist philosophies have included special principles held to be active in guiding and organising 'vital' processes. These types of philosophies, therefore, may be distinguished by the term 'vitalist'.

Of the various theorists who have employed a vital principle along with organicist interpretations of music, one writer has attracted particular attention in the musicological community. This writer is Heinrich Schenker (1868–1935), who was born near Podhayze in

[1] H. and E. H. Mueller von Asow (eds.), *The Collected Correspondence and Papers of Christoph Willibald Gluck* (New York, 1962), p. 14 (italics mine).

Galicia but in 1890 moved to Vienna, where he died.[2] From 1904 Schenker issued treatises on, and analyses of, music, all of which together constitute his theory of tonality.[3] This theory pertains to the musical 'language' that has governed most western art music from the seventeenth to the end of the nineteenth centuries, after which period another language—atonality—began to emerge. In his publications Schenker restricted himself to what he supposed were the 'masterworks' of tonality, namely, compositions written chiefly by Germans of the eighteenth and nineteenth centuries. His last treatise was published posthumously as *Der Freie Satz* (Vienna, 1935) and translated into English as *Free Composition* (New York, 1979), and it is this work I shall examine here, since it contains Schenker's most mature expression of his theory.[4]

---

[2] A. Forte and W. Drabkin, 'Heinrich Schenker', *The New Grove Dictionary of Music and Musicians* ed. by S. Sadie (20 vols., London, 1980), vol. 16, pp. 627–628.

[3] See L. Laskowski, *Heinrich Schenker: An Annotated Index to his Analyses of Musical Works* (New York, 1978); see also D. W. Beach, 'A Schenker bibliography', *Journal of Music Theory* (1969) 13: 2–37, (1979) 23: 275–286, who also includes writings concerning Schenker. Although the *name* of Heinrich Schenker is widely known in English-speaking musicological circles, investigations into his life and thought were impeded by a lack of availability of his original publications, the dense German style of his writings, a dearth of reliable translations of his works and the difficulty of access to two private collections of his manuscripts. Consequently, Schenker's theory of tonality has been disseminated chiefly through the writings and teachings of his pupils and disciples. Factionalism within the ranks of Schenker's followers, coupled with the provocative nature of some of Schenker's writings, too often resulted in polemics rather than in dispassionate, critical analysis of his thought. Fortunately, this state of affairs is changing: some of Schenker's writings are available in facsimile reprints; there are increasing numbers of translations of a more or less scholarly nature; and the private collections of his manuscripts are now housed in public repositories.

[4] All quotations herein are taken from H. Schenker, *Free Composition [Der freie Satz]* tr. and ed. by E. Oster (2 vols., New York, 1979). Oster based his translation on the second edition of *Der Freie Satz* (Vienna, 1956), which was edited and revised by Oswald Jonas. In the case of long quotations from Oster's translation, I provide the corresponding page numbers to the German second edition. Citations distinguish the two versions by referring either to Schenker/Oster or to Schenker/Jonas. However, both Oster and Jonas excised a number of passages from the first edition; and one of the editors of Oster's

Schenker's object is not to teach how to compose but to train students 'to hear music as the masters conceived it'.[5] Such an object poses a twofold problem, namely, to find an appropriate representation that accounts not only for the structure of music but also for the procedure of apprehending that structure. In what follows, Schenker's argument is analysed for the purposes of reconstructing his representation of music.[6] On the basis of this reconstruction, it is possible to place his work within the broader context of an intellectual tradition. The need for such an exercise has been well stated by Allan Janik and Stephen Toulmin as follows:

> [Those] who are ignorant of the context of ideas are ... destined to misunderstand them ... [for] the philosophical problems and ideas of actual men ... confront us like geological specimens in situ; and, in the process of chipping them free from their original locations, we can too easily forget the historical and cultural matrix in which they took shape, and end by imposing on them a sculptural form which reflects the preoccupations, not of their author, but of ourselves.[7]

Heinrich Schenker's Representation of Music

In *Free Composition* Schenker proposes a method of grasping how musical 'evolution' takes place not only in individual compositions but also in all compositions instancing tonality. His method, he

---

translation, John Rothgeb, re-inserted some of the passages in an appendix (Oster died before the initial stages of publication). These passages are cited as Schenker/Rothgeb.

[5] Schenker/Oster, p. xxii; see also p. xxiii, where Schenker states:

> Only by the patient development of a truly perceptive ear can one grow to understand the meaning of what the masters learned and experienced. If a student under firm discipline is brought to recognize and experience the laws of music, he will also grow to love them. He will perceive that the goal toward which he strives is so meaningful and noble that it will compensate for the fact that he himself may lack a genuine talent for composition.

[6] Schenker's theory is in process of formalisation; see M. Kassler, *Proving Musical Theorems I: The Middleground of Heinrich Schenker's Theory of Tonality* (Sydney, 1975).

[7] A. Janik and S. Toulmin, *Wittgenstein's Vienna* (London, 1974), pp. 27–28.

claims, incorporates three new teachings (*Lehren*): first, the concept of 'organic coherence' (*organisch Zusammenhang*); then, the concept of levels; and finally, the concept of music 'as a manifestation of the fundamental design'.[8] Organic coherence refers to the interrelatedness of all parts constituting a whole composition. This interrelatedness arises from, and is maintained by, the background level—one of three *Ursätze*, or primitive compositions, illustrated in *Figure 9.1*.[9] The upper notes of the *Ursätze*, to which Schenker gives the

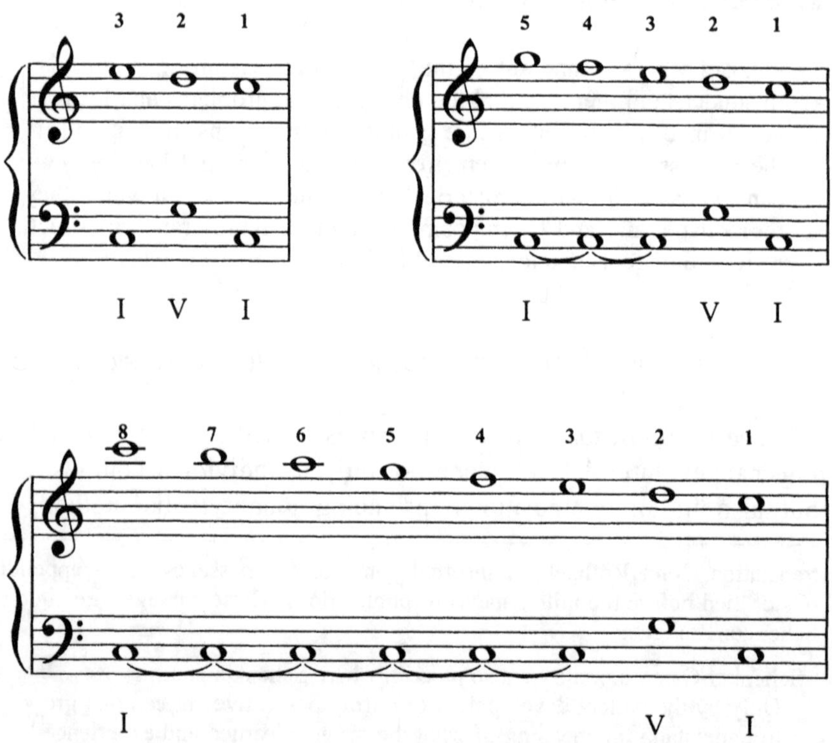

Figure 9.1. The three primitive compositions

---

[8] Schenker/Oster, pp. xxi, 3–4, 130.

[9] For convenience the key of C is employed; however, these primitive compositions may occur in any key.

name '*Urlinie*', and the lower notes are derived from the so-called 'chord of nature', that is, from the fundamental and first four notes of the overtone, or harmonic, series as shown in *Figure 9.2*.[10]

Figure 9.2. The chord of nature (so-called)

Schenker claims that the *Ursätze* manifest the chord of nature both as horizontal bass arpeggiation of the lower notes and as derived horizontal succession of the upper notes. He also claims that the upper notes, or *Urlinie*, define the 'tone-space' of compositions and that this tone-space is to be understood horizontally as duration. Since the *Urlinie* is the 'fountainhead' of musical form, its tone-space is indivisible and anterior to all form.[11] Hence, 'traversal' of the *Urlinie*

[10] According to Schenker/Oster, p. 10 (Schenker, Jonas, p. 39), the chord of nature in its vertical form

> cannot be transferred to the human larynx; nor is such a transfer desirable, for the mere duplication of nature cannot be the object of human endeavor. Therefore art manifests the principle of the harmonic series in a special way, one which still lets the chord of nature shine through. The overtone series, this vertical sound of nature, this chord in which all the tones sound at once, is transformed into a succession, a horizontal arpeggiation, which has the added advantage of lying within the range of the human voice. Thus the harmonic series is condensed, abbreviated for the purposes of art... .

Schenker's 'chord of nature' is not original with him but stems from late seventeenth- and early eighteenth-century attempts of music theorists to provide a 'natural' basis for the justification of harmonic practice, a pseudo-scientific tradition that now is referred to as 'natural-law' music theory. See B. L. Green, *The Harmonic Series from Mersenne to Rameau: A Historical Study of Circumstances leading to its Recognition and Application to Music* (Ohio State University doctoral dissertation, 1969), and M. Shirlaw, *The Theory of Harmony*...(2d edn., Dekalb, Illinois, 1955).

[11] Schenker/Oster, p. 16.

is 'the most basic of all passing-motions; it is the necessity (derived from strict counterpoint) of continuing in the same direction which creates coherence, and ... makes this traversal the beginning of all coherence in musical composition'.[12]

Musical content arises from the *Ursätze* by means of the 'confrontation and adjustment' of the upper notes (*Urlinie*) with the lower notes of the bass arpeggiation.[13] Confrontation and adjustment result not only in horizontalised versions of the chord of nature but also in forms representative of a state (*ein Zustand*) existing behind all compositions that instance tonality. Since a composition contains only one of these forms, that *Ursatz*, according to Schenker, 'represents the totality': 'It is the mark of unity and, since it is the only vantage point from which to view that unity, prevents all false and distorted conceptions'. In that one *Ursatz* resides 'the comprehensive perception, the resolution of all diversity into ultimate wholeness'.[14]

The *Ursatz* reveals 'the development of one single chord into a work of art'.[15] The key of this chord alone is present, and it is present from the start in the *Ursatz*. A musical composition is generated by the extension (*Dehnung*) of the one chord by means of what Schenker terms 'voice-leading levels' (*Stimmführungsschichten*) or 'transformations' (*Verwandlungen*). These are reached by applying various sequences of so-called 'prolongation techniques' that 'propagate' the form of the *Ursatz*.[16] Governed by the laws of counterpoint, the prolongation techniques extend a single chord from the level of background to that of middleground. This level consists of the large-scale structure of a composition—its principal sections and their tonal centres by which we recognise and identify the particular form of a composition—for example, sonata form or rondo form.

---

[12] Ibid., p. 12 (Schenker/Jonas, p. 41).

[13] Ibid., p. 13.

[14] Ibid., p. 5 (Schenker/Jonas, p. 28).

[15] Ibid., p. 112.

[16] Ibid., p. 87.

According to Schenker, it is impossible to generalise regarding the number of transformations necessary to reach the middleground, 'although in each individual instance the number can be specified exactly'. But 'the first two levels already contain the branching-out into the particulars of a work of art' and 'there are some prolongations which would occur only at the first level, others which take place only at the second'.[17] Extension of a chord does not cease with the middleground, but continues into the foreground, where, by another sequence of prolongation techniques, the harmonic, melodic and rhythmic details are specified. The foreground is the composition itself—for example, the first movement of Beethoven's *Fifth Symphony*, so that it distinguishes compositions as individuals, or specific pieces of music.

Schenker provides a special graphic means of representing the logical relationships between simple tone-successions and more complex ones, for an example of which, see the Appendix at the end of this chapter. These graphs (*Urlinie-Tafeln*) are not merely practical or educational aids that facilitate an understanding of his theory. Rather, they have 'the same power and conviction as the visual aspect of the printed composition itself (the foreground)', since Schenker intended his graphs as approximations of actual compositions.[18] Consequently, they define a system of sequences of empirical events by reference to a multidimensional tone-space, the coordinates of which represent all the independent variables in Schenker's theory. But Schenker cautions that even the 'most successful graphic representation of the logical relationships between background and foreground must fail to portray the ultimate reality'—that the *Ursatz* is

> always creating, always present and active; this 'continual present' in the vision of the composer is certainly not a greater wonder than that which issues from the true experiencing of a moment of time; in this most brief space we feel something very like the composer's perception, that is, the meeting of past, present, and future.[19]

---

[17] Ibid., p. 26 (Schenker/Jonas, p. 58).

[18] Ibid., p. xxiii.

[19] Ibid., p. 18 (Schenker/Jonas, pp. 49–50).

Consequently, 'the projection of the horizontal ... alone is the purpose and content of music'.[20] But there can be no content without a goal, for 'the goal and the course to the goal' come before content.[21] This is so, because the *Ursatz* 'signifies movement toward a specific goal', which is twofold.[22] First, there is the unfolding of the composition itself by means of 'progressive contrapuntal differentiation', that is, by various transformation levels from background to middleground to foreground.[23] All musical differentiation arises from the differences in the transformation levels that, in turn, 'lead to differences of form'.[24] Second, there is organic coherence, which is the highest goal of music, since only by means of organic coherence does music 'drive toward the organic human soul'.[25]

Music's dynamic drive for extension, however, is fraught with tension (*Spannung*), because

> In the art of music, as in life, motion toward the goal encounters obstacles, reverses, disappointments, and involves great distances, detours, expansions, interpolations, and, in short, retardations of all kinds. Therein lies the source of all artistic delaying, from which the creative mind can derive content that is ever new. Thus we hear in the middleground and foreground an almost dramatic course of events.[26]

For Schenker, therefore, the 'principle of inner tension and its corresponding outward fulfilment' is 'the highest principle which is common to all arts'.[27] This principle, which manifests itself

---

[20] Ibid., p. 117.

[21] Ibid., p. 5.

[22] Ibid., p. 4. The goal is movement to the upper fifth, 'and the completion of the course' takes place 'with the return to the fundamental tone'.

[23] Ibid., p. 15.

[24] Ibid., p. 131.

[25] Ibid., p. xxiv.

[26] Ibid., p. 5 (Schenker/Jonas, p. 49).

[27] Ibid., p. xxiv, where Schenker also states that man

differently in different material, is realised in music through specific compositional procedures. But in 'its linear progressions and other comparable tonal events, music mirrors the human soul in all its metamorphoses and moods'.[28] Hence,

> As the image of our life-motion, music can approach a state of objectivity, never, of course, to the extent that it need abandon its own specific nature as an art. Thus, it may almost evoke pictures or seem to be endowed with references, and connectives; it may use repetitions of the same tonal succession to express different meanings; it may simulate expectation, preparation, surprise, disappointment, patience, impatience, and humor. Because these comparisons are of a biological nature, and are generated organically, music is never comparable to mathematics or to architecture, but only to language... .[29]

Music does not express particular emotions or other things; rather, it suggests these things to us, since in the linear progressions 'the composer lives his own life as well as that of the linear progressions'. Consequently, 'their life must be his, if they are to signify life to us'.[30]

In a number of places in *Free Composition*, Schenker digresses to consider facets of the history of music (phylogeny). In one of these digressions he states that music did not always have a dynamic drive toward its goal, since in earlier times the 'word alone was the generator of tone successions'.[31] As long as the word 'interpreted music and determined its dimension', there was no progress in music, since music was 'absolved from the obligation to interpret and develop itself'.[32] But when music conformed to nature, the resulting advance—counterpoint—occurred, because music 'yearned for

---

lives his whole life in a state of tension. Rarely does he experience fulfilment; art alone bestows on him fulfilment, but only through selection and condensation.

[28] Ibid., p. xxiii.

[29] Ibid., p. 5 (Schenker/Jonas removed this passage).

[30] Ibid.

[31] Ibid., p. 93.

[32] Ibid., p. 99.

greater length, further extension in time, greater expansion of content from within, as do all physical or spiritual beings that obey nature's law of growth'.[33]

Only by means of counterpoint could music's vertical and rhythmic dimensions be clarified; only by means of counterpoint could the horizontal dimension be defined; only by means of counterpoint could 'a tonal succession ... achieve a specific inner relatedness, limited and meaningful, since it was based upon specific intervals and also specifically determined in the manifold time-values of the individual notes'.[34] More importantly, only the unity provided by counterpoint could lead to repetition, which, for Schenker, is 'a biological law of life, physical life as well as spiritual'.[35] But repetition signifies neither duplication nor imitation;[36] rather, it denotes movement from one transformation level to another, because

> As they move toward the foreground, the transformation levels are actually the bearers of developments and are, at the same time, repetitions or parallelisms in the most elevated sense—if we permit ourselves to use the word 'repetition' to describe the movement from transformation level to transformation level. The mysterious concealment of such repetitions is an almost biological means of protection: repetitions thrive better in secret than in the full light of consciousness... .[37]

Repetition, then, is 'a symbol of organic life in the world of tones, as though statement and variant were connected by bonds of blood'.[38]

---

[33] Ibid., p. 94.

[34] Ibid., p. 99.

[35] Ibid., p. 118. In Schenker's usage, repetition is the idea that development proceeds from the general to the special. Deriving from the theory of repetition in biology—that the earliest embryonic stages of related organisms are identical and that distinguishing features are added later as heterogeneity differentiates from homogeneity, this theory differs from that of recapitulation—the repetition of ancestral adult stages in embryonic or juvenile stages of descendants.

[36] Ibid., p. 159: 'Imitation is no substitution for evolution'

[37] Ibid., p. 18 (Schenker/Jonas, p. 50).

[38] Ibid., p. 99.

The motto of Schenker's work is: 'always the same, but not in the same way'.[39] This motto refers to Schenker's belief that all the masterworks of tonality manifest 'identical laws of coherence'. Hence, there is 'but one grammar of the linear progressions', namely, the grammar described by him 'in connection with the theory of coherence in music'.[40] Sameness of laws between masterworks in no way restricts the diversity—or the length—of compositions, but it does limit music to tonality. Thus, Schenker likens the search for new laws or new grammars of music to a 'quest for a homunculus', for he holds that the laws of musical coherence are compatible with nature:

> Music is not only an *object* of theoretical consideration. It is *subject*, just as we ourselves are subject. Even the octave, fifth, and third of the harmonic series are a product of the organic activity of the tone as subject, just as the urges of the human being are organic.[41]

Schenker therefore asserts that 'nature will endure, indeed, will conquer, in music also; she has revealed herself in the works of the masters and, in this form, she will prevail'.[42]

Since Schenker holds there can be no 'new way' of counterpoint, it may well be asked, in what sense is compositional activity free? The answer to this question is to be found in Schenker's belief that freedom lies only in creativity activity; and this activity is exercised, through the imagination, in choosing prolongation techniques. According to Schenker, a particular form of the *Ursätze* does not require or necessitate particular prolongations; if it did, all forms of the *Ursätze* would have to lead to the same 'prolongational forms'. Therefore, 'the choice of prolongations remains essentially free, provided that the indivisibility and connection of all relationships are assured'.[43]

---

[39] Ibid., title-page, pp. 5–6, and in other writings not considered here.

[40] Schenker/Rothgeb, p. 160.

[41] Schenker/Oster, p. 9 (Schenker/Jonas, p. 36).

[42] Ibid., p. 9.

[43] Ibid., p. 25 (Schenker/Jonas, p. 57).

The prerequisite for creative activity, however, is an impulse conceived as 'a vital natural power' (*einer lebendingen Naturkraft*) and which Schenker identifies as the emotion of love, for

> music, as art, has no practical benefit to offer. Thus there is no external stimulus for expansion of the powers of musical creativity and music's artistic means. The expansion of creative vision, then, must spring from within itself, only from the special form of coherence that is proper to it, and the special love intrinsic to it.
>
> Therefore the person whose tonal sense is not sufficiently mature to bind tones together into linear progressions and to derive from them further linear progressions, clearly lacks musical vision and the love that procreates. Only living love composes, makes possible the linear progressions and coherence—not metaphysics, so often invoked in the present time, or the much touted 'objectivity'; these, in particular, have neither creativity nor breeding warmth.[44]

The vital impulse, love, perseveres and is prolonged by will, by imagination and by inner necessity. By the term 'will' Schenker denotes music's tendency toward becoming a self-contained organism—toward living its own existence, for the culmination of music, both historically and compositionally, is to be found in music's striving for 'a likeness of itself, without having recourse to outside associations'.[45] Thus, while the *Ursatz* 'shows us how the chord of nature comes to life through a vital natural power', the 'primary power of this established motion must grow and live its own life'. This is so, because Schenker believes that 'that which is born to life strives to fulfill itself with the power of nature'.[46]

By the terms 'imagination' and 'inner necessity' Schenker denotes, on the one hand, creative activity exercised in choosing prolongation techniques to propagate one of the forms of the *Ursätze* and, on the other hand, the transformation levels that assure outward motion. Thus a tension is set up between imagination as freedom and inner necessity as determinism—'musical causality' or the inner logic of music that arises from the laws of counterpoint. Just 'as life is an

---

[44] Schenker/Rothgeb, p. 160 (Schenker/Jonas, p. 31).

[45] Schenker/Oster, p. 93.

[46] Ibid., p. 25.

uninterrupted process of energy transformation, so the voice-leading strata represent an energy transformation in the life which originates' in the *Ursatz*.[47] Indeed, Schenker suggests that music has a will of its own, which stems not only from its drive toward organic coherence but also from the inner necessity of the transformations.[48] The only necessity of these transformations is completion of certain types of motion, up or down. Each transformation is independent of the others and 'brings its own nature and purpose to fulfillment'; however, all transformations, taken together, form a self-contained unity, the final goal of which is the foreground or composition itself.[49]

For Schenker, then, the only reality we know is that of our own experience. But experience is to be understood not merely as that of the individual but also as that of a tradition, since

> The origin of every life whether of nation, clan, or individual, becomes its destiny... .
>
> The inner law of origin accompanies all development and is ultimately part of the present.
>
> Origin, development, and present I call background, middleground, and foreground; their union expresses the oneness of an individual, self-contained life.
>
> In the secret perception of the interaction of origin, development, and present, as well as in the cultivation of this awareness until it becomes definite knowledge, lies what we call tradition: the conscious handing down, passing on of all relatedness which flows together into the wholeness of life.
>
> To the person who is vitally aware of such relatedness, an idea is also part of real life, be that idea, religion, art, science, law, the state. Therefore the principle of origin, development, and present as background, middleground, and foreground applies also to the life of the idea within us.

---

[47] Schenker/Rothgeb, p. 160.

[48] E.g., Schenker/Oster, p. 44: 'That which is both will and necessity in the fundamental line [*Urlinie*] is also will and necessity in the derived linear progressions, in other words, the derived linear progression wants itself to be a true linear progression', namely, one that instances the 'eternal shape of life'. (Schenker's first music periodical was entitled *Der Tonwille*.)

[49] Ibid., p. 52.

> In order to comprehend what lives and moves behind the phenomena of life, behind ideas in general and art in particular, we ourselves require a definite background, a soul predisposed to accept the background. Such a soul, which constitutes a peculiar enhancement of nature in man—being almost more art than nature—is given only to genius.[50]

Experience pertains to the life of the body—the general sense or feeling of existence arising from the sum of bodily impressions, as well as to the life of the mind—the definite sensations of the special senses. Both kinds of life, according to Schenker, may be known from reflection, even though both kinds of life are inseparable if there is to be any life at all. By reflecting, then, upon our experience, we discover a unity or agreement between nature and art in that both are organic and both obey nature's law of growth. Indeed, growth in music is analogous to the growth in our own bodies, for

> It should have been evident long ago that the same principle applies both to a musical organism and to the human body: it grows outward from within. Therefore it would be fruitless as well as incorrect to attempt to draw conclusions about the organism from its epidermis.
>
> The hands, legs, and ears of the human body do not begin to grow after birth; they are present at the time of birth. Similarly, in a composition, a limb which was not somehow born with the middleground and background cannot grow to be a diminution [that is, an embelllishment of the foreground].[51]

But music obeys nature's law of growth through artistic means—the 'special laws' of counterpoint that govern the growth of compositions instancing tonality and mould them into self-contained organisms. Music's chief parameters, therefore, are duration and becoming; and, for Schenker, becoming is movement toward a goal. With the notion of a goal, of course, we encounter a teleonomic principle, that is, an idea of an oriented, coherent and constructive activity. This activity is creative activity, which is not finalistic in the sense that music tends toward the completion of some pre-established design. Rather, the tendency is given at the beginning through the

---

[50] Ibid., p. 3 (Schenker/Jonas, p. 25).

[51] Ibid., p. 6 (Schenker/Jonas, p. 31).

vital, natural power of love, the only 'end' of a piece of music being growth to maturity.

Schenker, therefore, conceives music as a symbol of living experience, for it not only mirrors the life of those who created masterworks but also suggests that life to those who would comprehend the masterworks. Comprehension, for Schenker, arises from intuition conceived as a sympathetic relationship between music as object and listener as subject:

> As a motion through several levels, as a connection between two mentally and spatially separated points, every relationship represents a path which is as real as any we 'traverse' with our feet. Therefore, a relationship actually is to be 'traversed' in thought—but this must involve actual time.[52]

In our traversal toward comprehension, Schenker holds that 'the standard for judging evolutionary plateaux derives from art as pure idea—whoever has fathomed its secrets knows that such an idea remains ever the same, ever indestructible, as an element of an eternal order'.[53]

The pure idea, which Schenker claims to have discovered, is the *Ursätze*, for these primitive compositions enable musical experience to be lifted to the realm of religious experience: the 'highest art of the genius takes part in human life as they themselves live it, and ... this high art furthers life and health just as milk and bread do, and can lead to Eros in the way any sacrament does'.[54] Thus, Schenker believes not only that life is parallel to, and inseparable from, music but also that this organic whole embraces the divinity as an immanent creative principle, for, after quoting Genesis 1.2. that 'the spirit of God moved upon the face of the waters', he asserts: 'the Creative Will has not yet been extinguished. Its fire

---

[52] Ibid., p. 6 (Schenker/Jonas, p. 32).

[53] Schenker/Rothgeb, p. 161 (Schenker/Jonas removed this passage).

[54] Schenker/Oster, p. 4, perhaps from Plato, *Symposium* (196e2–3), who paraphrased Euripides that possession by Eros makes every man a poet and who, in the same work (201d8 ff.), described the higher, intellectual Eros as a desire of generation in the beautiful (where the resulting poem is the poet's progeny).

continues in the ideas which men of genius bring to fruition for the inspiration and elevation of mankind'.[55] For Schenker, then, the life of physical and psychological phenomena are both manifestations of an ultimate reality that cannot be identified either with matter or with mind. This ultimate reality is love (God), the vital, procreative principle that is the spiritual cause of creation in the universe and that gives unity to the organised world.

From the foregoing analysis of Schenker's argument, we may conclude that a composition never is at any moment; it is always becoming. To construct a representation of music as process, Schenker must symbolise the three central tenets of his theory of music. First, there is his conception of a compositional whole as a set of parts differing one from another. Since, for Schenker, these differences depict a form, he symbolises a musical whole as a harmony of differences. Second, there is his notion of development, which takes place in time as well as in the repetition of processes. Any finite process must be completed before we can apprehend the piece as a whole, so that the commencement of a composition cannot reveal its full nature. To understand the process of compositional development, it is necessary to know the state of a composition's complete development (the foreground). Schenker thus defines development in terms of a composition's maturity; and he conceives development as a series of stages that a composition passes through on its progress to maturity. To symbolise this conception, Schenker represents the developmental processes as reproduction with variation.

Third, there is Schenker's notion of teleology, according to which a compositional whole (the foreground) is the end of organic growth. Given the complete process, we may analyse it into a series of stages that succeed one another in time; and we may note that each stage is necessarily different from every other. But the process of growth requires also that we notice what is essential to each stage, namely, that it develops into the next by its very nature and cannot develop into any other. Nor can there be a development from the first stage to the last that does not pass through all the intervening stages.

---

[55] Ibid., p. xxiv (Schenker/Jonas, p. 22).

To represent teleology, therefore, Schenker requires the idea of potentiality, and this idea is to be found in his concept of the *Ursatz*. The *Ursatz* 'is' only insofar as it is active, and its activity consists in a continuous transformation from one new state to another as it produces these states out of itself in an unceasing succession. Thus, each of the three *Ursätze* contains its own past (background) and is pregnant with its future (foreground). Hence, the compositional whole to be grasped cannot be symbolised mechanistically as a sum of its parts. Instead, the whole must be, and is, symbolised as presupposed by its parts and as constituting the condition of the possibility of their nature and being.

In constructing this kind of representation of music, Schenker's chief concerns are unity in multiplicity, coordination and differentiation of parts, constancy in change, being in becoming. These concerns proceed from Schenker's aesthetic apprehension of a musical organism as a whole, in which the unity of the whole is maintained by a harmony of differences. These differences are symbolised as differences of function in a unitary process that is the 'life' of the composition as a whole. The potentiality for life is given at the beginning, but it is not realised until the end of the growth process. Analysis of this type of symbolic construction is by function, one function revealing itself as demanding the next by an inner necessity. Consequently, change is represented as a differentiation of functions in maintaining the developing whole, whereas constancy is symbolised as the unity of form and function that are treated as inseparable.

It is to be noted, however, that Schenker's representation does not explain why musical growth takes place but only describes the fact. Indeed, Schenker states that 'I would not presume to say how inspiration comes upon the genius, to declare with any certainty which part of the middleground or foreground first presents itself to his imagination: the ultimate secrets will always remain inaccessible to us'.[56] Instead, his attention is directed to the growth of compositional complexity, which develops gradually from simple or primitive compositions (*Ursätze*). Thus, even though all 'forms appear in the ultimate foreground', 'all of them have their origin in,

[56] Ibid., p. 9.

and derive from the background'.[57] But they grow to maturity because of an internal force or directed entelechy working through the primitive compositions. In short, Schenker conception of 'evolution' is preformationist, ontogenetic and vitalistic.[58]

The Historical Matrix of Schenker's Theory

For his philosophy of musical 'evolution', Schenker is indebted to the work of two earlier writers: Johann Wolfgang von Goethe (1749–1832) and Arthur Schopenhauer (1788–1860).[59] Goethe held that species do not evolve but are created and that generation leads only to transformation within one type of species. For him, transformation, or the creation of new natural forms (variation), resulted not from chance but from the effort of the living creature itself to adapt to the

---

[57] Ibid., p. 130.

[58] In biology, preformationism is the theory that all parts of the organism pre-exist in the germ and that these pre-existing parts merely increase in size in the process of reproduction. The opposing theory, epigeneticism, holds that morphological complexity develops gradually during embryology from simple beginnings in an essentially formless egg (for the first modern version of this theory, see Chapter 4 *supra*). The term 'ontogeny' refers to the life history of an individual, both embryonic and postnatal, whereas, after the work of Charles Darwin, the term 'evolution' commonly refers to phylogeny—the life history of the species. Schenker, however, employs the term 'evolution' to denote not only the life history of individual compositions but also of all compositions instancing tonality. For him, therefore, ontogenetic 'evolution' is the basis of phylogenetic evolution. The implication is orthogenetic—that evolution once started in certain directions cannot deviate from its course (even though it may lead a lineage to extinction), since musical species are created and the term and 'end' of their evolution is tonality. For vitalism in biology, see L. von Bertalanffy, *An Introduction to Theoretical Biology* tr. by J. Woodger (New York, 1962), first issued as *Kritische Theorie der Formbildung* (Berlin, 1928).

[59] For the relation of their thought to that of the rationalist philosopher, Benedict de Spinoza, and the mystic, Jacob Boehme, see, e.g., E. Cassirer, *Rousseau, Kant, Goethe*...(Princeton, 1945); R. D. Gray, *Goethe the Alchemist*...(Cambridge, 1952); and A. Schopenhauer, *The World as Will and Representation* tr. by E. F. J. Payne (2 vols., New York, 1966), the concluding pages of which state the relation in which Schopenhauer's teaching stands to Spinozism in particular and pantheism in general.

circumstances of its existence.[60] Goethe's theory was intended as a means of explaining the origin and inner development of natural forms, where natural forms were defined as the interactions of systems of component parts, the development of which, through various stages, is from a seed or embryo. In his approach to form, Goethe rejected the mechanistic conception of the relationship of parts as mere succession and coexistence, one part being affected by its immediate neighbours. Indeed, he asserted that no examination of a part in isolation or of the external proportions could give a complete account of the whole, because outward shape is inseparable from inner patterning.

But Goethe believed that special principles were active in guiding and organising natural processes, and he sought to discover these principles by means of a formula, namely, when any definite or immutable state is presented to a vital process, it resists that state. Thus, inspiration presupposes expiration, systole presupposes diastole, contraction presupposes expansion. In 1790, for example, he employed the special principle of contraction and expansion to explain the metamorphoses of plants:

> When the plant vegetates, blooms, or fructifies, so it is still the same organs which, with different destinies and under protean shapes, fulfil the part prescribed by Nature. The same organ which on the stem expands itself as a leaf, and assumes a great variety of forms, then contracts in the calyx—expands again in the corolla—contracts in the reproductive organs—and for the last time expands as the fruit.[61]

According to Goethe, the plant, like all natural phenomena, expresses a life force; and this life force is none other than God's creative will,

---

[60] For Goethe's theory of ontogeny ('evolution') as the unfolding of form, see Gray, *Goethe the Alchemist* (note 58), pp. 71–100; G. A. Wells, *Goethe and the Development of Science, 1750–1900* (Alphen aan de Rijn, 1978), pp. 27–46; and E. M. Wilkinson, 'Goethe's conception of form', *Proceedings of the British Academy* (1951), 57: 175–197.

[61] J. W. von Goethe, *Versuch die Metamorphose der Pflanzen zu erklaren* (1790), tr. by A. Arber, 'Goethe's botany', *Chronica Botanica* (1946), 10: 1–126, p. 114. Goethe also utilised the formula in his *Farbenlehre* (1810), for which see R. Matthaei (ed.), *Goethe's Colour Theory*...tr. by H. Aach...(London, 1971), p. 256.

which is free. As a mirror of God's thought, nature is also free, since objects in nature exist by the mere necessity of their own inner character and are determined in their actions by themselves alone. God and nature are therefore a unity, which Goethe conceived as an artist who innovates, not merely imitates and who manifests artistic activity in the process of creation. Hence, Goethe could argue that the habitual distinction between form and function has no reality, since function is merely form thought of as activity. He could also argue that the standards for judging the work of nature are not absolute norms, external to the work, but laws inherent within the work, relative and accessible only through genetic analysis.[62]

To the study of forms that he developed, Goethe gave the name 'morphology'. By this term he denoted an independent science of change and transformation, not merely a branch of botany or zoology. He defined the work of the morphologist as apprehension of the full complexity of a living organism by means of intuitive contemplation (*Anschauung*) consisting of three activities: a constant alternation between analysis and synthesis; a sudden flash of insight or '*aperçu*'; and the employment of archetypal ideas.[63] Goethe regarded the last activity as a sustained process of moving freely between deduction and induction while comparing individual forms until the mind is sufficiently saturated with them for an archetype to emerge. Although these archetypes are abstractions, or what Goethe called '*Urphänomene*', they are not solely mental abstractions, for they partake of sensuous experience. Nor are the archetypes static, like Platonic Ideas, for Goethe held that our conceptions of archetypes are capable of modification as new forms are investigated.

Goethe regarded archetypal ideas as permanent, eternal manifestations of God's thought that underlie nature's ceaseless flux.

---

[62] For this kind of analysis, see Chapter 4 note 22 *supra*.

[63] For Goethe's method, see F. Heinemann, 'Goethe's phenomenological method', *Philosophy* (1934), 11: 67–81; W. Heisenberg, 'Goethe's view of nature and the world of science and technology [1967]', *Across the Frontiers* tr. by P. Heath (New York, 1974), pp. 122–141; H. von Helmholtz, 'On Goethe's scientific researches [1853]', *Popular Scientific* Lectures...(New York, 1962), pp. 1–21; H. Spinner, *Goethes Typusbegriff* (Zürich, 1933); Wells, *Goethe and the Development of Science* (note 59); and Wilkinson, 'Goethe's conception of form' (note 59)

To discover the *Urphänomene*, therefore, Goethe recommended two methods: vision and prolonged study. If one could visualise the archetype, that is, the idea in the mind of God/nature, it would be possible to grasp the essential character of individual forms, for these forms, according to Goethe, are but modifications—metamorphoses—of the respective archetypes. If one could conduct a prolonged study of existing forms, careful scrutiny would reveal their 'formal essence', the archetypal characters which appear disguised in any individual forms. Thus, Goethe wrote to a friend:

> When you say that one can only believe in God ... I reply that I rely more on seeing. And when Spinoza speaks of 'intuitive knowledge,' and says: 'This species of knowing proceeds from an adequate idea of the formal essence of certain attributes of God to the adequate knowledge of the essence of things': these few words give me the courage to devote my whole life to the contemplation of things that are within my reach, and of whose formal essence I can hope to construct an adequate idea.[64]

In whatever manner the study of forms proceeded, the principal task of the morphologist, Goethe held, was to find the *Urphänomene*.[65]

It now is known that Schenker's life-long study of the German 'masterworks' also proceeded through intuitive contemplation—through repeated listening, reading and playing scores, through analysis and graphing of music. By means of these activities he claims to have discovered the *Ursätze* and the manner of their unfolding. Schenker's *Ursatz* is an archetypal or 'pure' idea which gives rise to his motto, 'always the same but not in the same way'.

---

[64] Letter dated 5 May 1786, quoted in Wells, *Goethe and the Development of Science* (note 59), pp. 23–24.

[65] Quoted in Wells, ibid. (note 59), p. 30. About the *Urpflanze* or ideal plant type, Goethe wrote that

> With this model and the key to it … one can invent plants *ad infinitum*, and they will be consistent, that is they could exist even if in fact they do not; they are not poetic shades or phantoms, but possess an inner truth and necessity. The same law will be found applicable to all other living things.

Thus, Goethe's *Urphänomene* are not common ancestors of individual things; rather, they are archetypal ideas of which all things, existing or not, are modifications.

This motto is reminiscent of Goethe's conception of unity, a conception that was articulated by him in relation to music, when he wrote to a friend:

> At present Zelter is here, and through his presence, I am probably progressing in my desire to extract something of use from music theory, from my point of view, in order to connect it with the rest of physical science and also with the theory of colours. If a few great formulas come off, then *everything must become one, everything must spring from one and return to one.*[66]

In addition to a prolonged study of nature, however, Goethe believed that *Urphänomene* could be discovered by vision. A century later Schenker echoes this belief in his assertion that:

> Inasmuch as all religious experience, and all branches of philosophy and science press for the shortest formulae, a similar urge led me to conceive also ... [musical] composition not only out of the nucleus of the *Ursatz* at the first composing out of the fundamental chord ... I was given a vision of the *Urlinie*, I did not invent it![67]

The search for a single formula, archetype or *Urphänomene*, then, is as basic to Schenker's thought as it was to Goethe's. Moreover, Schenker shares with Goethe a concern to explain the origin and dynamic inner development of forms ontogenetically. To accomplish these tasks, Goethe had devised a method that Schenker adopts and advocates, as is evident in Part I, Chapter 1 of *Free Composition*, where he quotes Goethe as follows:

---

[66] J. W. von Goethe, *Briefe der Jahre 1786–1814*...(Zürich, 1962), p. 610 (italics mine), dated 19 July 1810. For some of the results of Goethe's investigations with Carl Friedrich Zelter, see A. D. Coleridge (tr.), *Goethe's Letters to Zelter, with Extracts from Those of Zelter to Goethe* (London, 1887), pp. 267–276 *et passim*. Zelter was a pupil of the music theorist, Johann Philipp Kirnberger, who in turn had been a pupil of Johann Sebastian Bach.

[67] S. S. Kalib (tr.), *Thirteen Essays from the Three Yearbooks 'Das Meisterwerk in der Musik' by Heinrich Schenker*...(3 vols., Northwestern University doctoral dissertation, 1973), vol. 2, p. 218 (I have emended the germanised English of the translator).

> Sometimes a most curious demand is made: that one should present experience and perceptions without recourse to any kind of theoretical framework, leading the student to establish his conviction as he will. But this demand cannot be fulfilled even by those who make it. For we never benefit from merely looking at an object. Looking becomes considering, considering becomes reflecting, reflecting becomes connecting. Thus, one can say that with every intent glance at the world we theorize. To execute this, to plan it consciously, with self-knowledge, with freedom, and, to use a daring word—with irony—requires a considerable degree of skill, particularly if the abstraction which we fear is to be harmless and if the empirical result which we hope to achieve is to be alive and useful.[68]

For Goethe, the observed and the theoretical were not opposites, because he considered every observation as being itself theoretical. Thus, he argued that the responsibility for understanding nature rests with us, but our understanding arises from a dialogue with nature. To understand her language, we have to have flexible and formative minds, for if there is no form within us, we shall not find it outside us. The discovery of forms, Goethe insisted, is a matter of choice no less than a matter of understanding. If we do not discover them, the loss is ours; for where there is no form, there is no meaning.[69] These sentiments resound throughout Schenker's writings.

It is notable, however, that Goethe never solved a particular problem in morphology, namely, how to symbolise his findings. He insisted that morphology, as the science of change and transformation, could not rely on mathematical symbols. For him, such symbols were static and reductionist and, hence, incapable of representing form as a dynamic, developing system of relationships. In attempting to solve the problem of symbolic representation, Goethe relied on drawings, on discursive language and on poetic statement. However, the problem remained: 'how could one find a symbolism for expressing simultaneously process and permanence, the unceasing transformation of nature and her tendency to persist in

---

[68] Schenker/Oster, p. 3, from Goethe's *Farbenlehre* (1810).

[69] Wilkinson, 'Goethe's conception of form' (note 60), pp. 189–190.

specific forms?'[70] According to Goethe, a formulation that approximates as closely as possible to reality should be able to express both process and permanence at once. For this, he wrote in 1823, a 'symbolism would have to be created', adding: 'But who is to achieve this? Who is to acknowledge it after it has been done?'[71]

A step toward the creation of symbols for representing both permanence and flux was taken by Goethe's younger contemporary and friend, Arthur Schopenhauer, who adopted Goethe's conception of nature as a creative artist and asserted that

> each creature is its own creation. Nature, who never lies and is as unsophisticated as genius, frankly expresses the same thing, for each being only lights, as it were, its own torch at that of another, its exact replica, and then before our very eyes makes itself, taking its material from outside, but its form and movement from within itself; this we call growth and development. Thus, empirically, every creature stands before us as its own creation. But men do not understand the language of Nature, because it is too simple.[72]

Like Goethe, Schopenhauer assigned an important role to intuition, regarded duration and movement as principal modes of reality and held that physical and psychological phenomena alike are manifestations of an ultimate reality which cannot be identified either with mind or with matter. But Schopenhauer's ultimate reality—the will to live—differs from Goethe's, for it is neither divine nor rational. Indeed, Schopenhauer argued that sleep revealed to us the true primacy of the irrational will, for in sleep the intellect withdraws while the will remains active—witness the motion of the heart which 'alone is untiring, because its beating and the circulation of the blood are not conditioned directly by the nerves, but are just the original expression of the will'.[73]

---

[70] Ibid.

[71] B. Mueller (tr.), *Goethe's Botanical Writings*...(Honolulu, 1952), p. 117.

[72] A. Schopenhauer, 'Vergleichende Anatomie', *Sämmtliche* Werke ed. by J. Frauenstädt (6 vols., Leipzig, 1919), vol. 4, p. 28, quoted in E. S. Russell, 'Schopenhauer's contribution to biological theory', *Science, Medicine and History*...ed. by E. A. Underwood (2 vols., London, 1953), p. 209.

[73] Schopenhauer, *The World as Will* (note 59), vol. 1, p. 240.

Schopenhauer described the separation of will from intellect in terms of the heart's diastole (expansion) and systole (contraction): the intellect, like the hard and difficult diastole, is for the organism merely a means, whereas the will, like the beneficent systole, is for the organism its end. Schopenhauer's will, then, is an unconscious, striving, irrational power that 'objectifies' itself in the phenomenal world. In the process of unfolding, each individual thing embodies the will to live, and differentiation of types and increasing complexity in the phenomenal world are due to the striving of the will for maximum expression. Each individual thing, as an objectification of the one will to live, strives to assert its own existence at the expense of other things. Hence, the world is a field of conflict, which manifests the nature of the will at variance with itself.

These ideas are systematically expounded in Schopenhauer's major treatise, *The World as Will and Representation*, published first in 1819 and expanded in a second edition of 1844. Here, the author conceived music as 'an unconscious exercise in metaphysics in which the mind does not know it is philosophizing',[74] because music is related to the will 'as the depiction to the thing depicted, as the copy to the original'.[75] For Schopenhauer, however, music is 'directly a copy of the will itself, and therefore [expresses] the metaphysical to everything physical in the world'. Hence, he supposed that 'we could just as well call the world embodied music as embodied will', thereby suggesting that music is the key to metaphysical reality.[76] But music is related to physical reality as well, since Schopenhauer believed that natural objects are classed according to a hierarchy, scale or series of

---

[74] Ibid., vol. 1, pp. 256, 264, a modification of Gottfried Wilhelm Leibniz, 'On the ultimate origin of things [1697]', *Selections* ed. by P. P. Wiener (New York, 1951), p. 532: 'Music charms us, although its beauty only consists in the harmonies of numbers and in the reckoning of the beats or vibrations of sounding bodies, which meet at certain intervals, reckonings of which we are not conscious and which the soul nevertheless does make'. This conception, that music is an unconscious exercise in arithmetic in which the mind does not know it is counting, may be traced to the influence of Thomas Hobbes. See J. C. Kassler, *Inner Music: Hobbes, Hooke and North on Internal Character* (London, 1995).

[75] Ibid., vol. 1, p. 256.

[76] Ibid., vol. 1, pp. 262–263.

gradations. He treated these gradations—but not the objects themselves—as eternal and immutable; that is, as similar to Platonic Ideas.

In its very structure music also exhibits gradations, and Schopenhauer argued that these gradations parallel the Platonic Ideas. In attempting to explain how this parallelism operates, he provided an outline of the will's objectification that begins with the lowest grades, namely, inorganic nature, 'the mass of the planet'. In music, the parallel to inorganic nature is the 'ground-bass', which Schopenhauer derived from the chord of nature.[77] The ground-bass parallels 'the crudest mass on which everything rests and from which everything originates and develops' in the world; that is, the ground-bass is analogous to

> the fact that all the bodies and organizations of nature must be regarded as having come into existence through gradual development out of the mass of the planet. This is both their supporter and their source, and the high notes have the same relation to the ground-bass. There is a limit to the depth, beyond which no sound is any longer audible. This corresponds to the fact that no matter is perceivable without form and quality, in other words, without the manifestation of a force incapable of further explanation, in which an Idea expresses itself, and, more generally, that no matter can be entirely without will.[78]

Between the ground-bass and the melody are various gradations of the Ideas in which the will objectifies itself. Those grades 'nearer to the bass are lower ... grades, namely, the still inorganic bodies manifesting themselves, however, in many ways'. Those grades that are higher represent to Schopenhauer the plant and animal worlds,

---

[77] Schopenhauer's 'ground-bass' could be interpreted according to one of two theories of harmony—that of Jean-Philippe Rameau or that of Johann Philipp Kirnberger. From the views expressed by Schopenhauer, ibid., vol. 2, p. 456, it seems likely that he had Kirnberger's theory in mind, a theory in which music may be reduced to, and derived from, two different classes of chords—essential and accidental. Essential chords are the triad and the chord of the seventh; accidental chords are those formed by suspension, anticipation and transition.

[78] Ibid., vol. 1, p. 258.

whereas the 'definite intervals of the scale are parallel to the definite grades of the will's objectification, the definite species in nature'.[79]

At the highest grade is melody, which Schopenhauer conceived as parallel to the intellectual life and endeavour of man. For him, only melody has 'intentional connexion from beginning to end'; consequently, it relates 'the story of the intellectually enlightened will, the copy or impression whereof in actual life is the series of its deeds'.[80] At this highest grade, however, music

> does not express this or that particular and definite pleasure, this or that affliction, pain, sorrow, horror ... but ... their essential nature, without any accessories, and so also without the motives for them. Nevertheless, we understand them perfectly in this extracted quintessence. Hence, it arises that our imagination is so easily stirred by music, and tries to shape that invisible, yet vividly aroused, spirit-world that speaks to us directly... .[81]

Indeed, only in music may we hear 'the secret history of our will and of all its stirrings and strivings with their many different delays, postponements, hindrances, and afflictions, even in the most sorrowful melodies'.[82]

In Schopenhauer's philosophy, then, the phenomenal world (nature) and music are merely two expressions of the same thing—the will; and 'this thing itself is therefore the only medium of their analogy, a knowledge of which is required if we are to understand that analogy'.[83] But Schopenhauer warned his readers that the many analogies that he brought forward in treating music had no direct relation to music, since music never expresses the phenomenal world but only the inner nature of that world, namely, the will of every phenomenon. Analogies, according to Schopenhauer, are simply a means of shaping our subjective experience or, more precisely, representing that experience in words:

---

[79] Ibid.

[80] Ibid., vol. 1, p. 259.

[81] Ibid., vol. 1, p. 261.

[82] Ibid., vol. 2, p. 451. Compare this with Schenker's statement above (note 26).

[83] Ibid., vol. 1, p. 262.

> I have devoted my mind entirely to the impression of music in its many different forms; and then I have returned again to reflection and to the train of my thought expounded in the present work, and have arrived at an explanation of the inner essence of music, and the nature of its imitative relation to the world, necessarily to be presupposed from analogy. This explanation is quite sufficient for me, and satisfactory for my investigation, and will be just as illuminating also to the man who has followed me thus far, and has agreed with my view of the world. I recognize, however, that it is essentially impossible to demonstrate this explanation, for it assumes and establishes a relation of music as a representation, and claims to regard music as the copy of an original that can itself never be directly represented.[84]

In his book, *Kontrapunkt*, on strict composition, Schenker appears to reject Schopenhauer's belief that music is a symbol of the will, or as Schenker states that belief, 'that music represents "the innermost core which comes before all form, the heart of things"'.[85] Such a belief, Schenker asserts,

> is precisely in contradiction with music ... thus one sees how, in spite of frequently right ideas, the lack of clarity of the philosopher ultimately brings about his downfall. Music is not the 'heart of things', indeed not, it does not want to have much or anything at all to do with things; tones are themselves, organisms, as it were, with their own social laws etc.
>
> And if he had first been able to comprehend and master the absolute [nature] of music from [the point of view] of counterpoint, how easy would it be then for the philosopher to comprehend all the better the ultimate secret of the world, its own absolute existence, the dream of the creator as a similarly absolute event![86]

But Schopenhauer never claimed that music symbolised things; rather, he believed that music represented the inner nature of the world, the will. This symbolism, but with a different interpretation,

---

[84] Ibid., vol. 1, pp. 256–257.

[85] H. Schenker, *Kontrapunkt, Erster Halbband*...(Stuttgart, 1910), pp. 23–24.

[86] Ibid.

was retained by Schenker, whose will is neither unconscious nor irrational, because it is guided by the conscious and rational power of love.

With this distinction in mind, we may say that Schenker's *Ursatz* is analogous to Schopenhauer's ground-bass, the lowest grade of the will's objectification in the phenomenal world. To reach the higher grades—the middleground and foreground—various transformations must take place to ensure that the musical organism grows to maturity. During these transformations the *Ursatz* remains indivisible, being eternally and wholly present in compositions instancing tonality, just as Schopenhauer's will or inner being is eternally present, whole and undivided, in all of nature. Thus, Schenker states that 'because of the sense-perceived animated movement of its innate horizontal spans, and because of the fact that it possesses allegorical appeal to the human soul through its movements', music 'may very well be considered the most independent and noblest of all the arts'.[87]

To explain the gradations of the will, Schopenhauer relied on a scale-of-nature analogy. This mode of explanation is too static for Schenker, who rejects it in favour of Goethe's dynamic theory of generation and transformation in order to show that music is an organism with a life of its own.[88] But his representation of music as unfolding life does not meet Goethe's requirement for symbolism. According to Ronald Gray, Goethe made a distinction between allegories, metaphors and symbols, holding that an allegory is a story or image that requires explanation. As in a simile, an allegory needs a link—'this is like that'. In a metaphor the link begins to disappear, so that we have 'this is that'. But in a symbol there no longer is any distinction at all.[89] Schenker's representation of music falls within the realm of Goethe's allegory and metaphor, for it is representation as *Vorstellung*, that is, as sensory or perceptual images which are distinct from the abstract conceptions derived from perception. This same type of representation is to be found in Schopenhauer's treatise, and it has nothing to do with Goethe's notion of symbol.

---

[87] Kalib (tr.), *Thirteen Essays* (note 67), vol. 2, p. 218.

[88] Ibid., vol. 2, pp. 511–512.

[89] Gray, *Goethe the Alchemist* (note 59), pp. 130–131.

For Goethe, symbol is representation as *Darstellung*—as consciously constructed schemes for knowing. Representation as *Darstellung* requires three things: that it has logical consistency and simplicity of presentation; that it corresponds with empirical data; and that the whole theory is present. Hence, representation as *Darstellung* denotes a model of a theory, and Schenker achieved this kind of representation in his graphing technique.[90] Although, on first consideration, Schenker's models or graphs may seem to simplify musical experience, the complexity of that experience is to be apprehended gradually by repeated graphing of the same composition. Even Schenker was dissatisfied with some of his early graphs, which he then altered in later publications. His graphs, therefore, are not static models but processes for striving toward the acquisition of holistic knowledge of a musical composition, as well as for grasping the dynamic causality of inner events by which compositions instancing tonality unfold according to Schenker's theory. Above all, the graphs provide a means of representing both the permanence of the *Ursatz* and the flux of the transformations. Schenker thus provided an original solution to Goethe's problem of symbolism.

CONCLUSION

With his vision of the musical universe as an organic and divine whole, Schenker stands firmly within the neo-Platonic tradition as represented by a long line of thinkers. A characteristic of this tradition is syncretism, so that we may expect to find neo-Aristotelian features in Schenker's work. Such features are indeed present in his notions that form cannot exist apart from substance, mind apart from matter; that matter is inchoate, the potential, characterised by a tendency to manifest itself in specific forms; that form is that which has actuality. But to describe Schenker's work as neo-Platonic tells us little about the main tenets of his philosophy of

---

[90] Schenker himself employs the term '*Darstellung*' to refer to his graphic representations; see Schenker/Jonas, pp. 62, 66, 68, 80, 85 *et passim*.

music. These tenets are: monism, duration, movement, entelechy and intuition.

Monism. Schenker's monism stems from his belief—shared by Goethe—that a benevolent, rational God (love) is the ultimate reality behind all phenomena. Love, then, is Schenker's metaphysical principle, for it is original and must be taken on faith. Schenker's position in this regard agrees with Schopenhauer, who stated that

> The real foundation of all truths which ... are called metaphysical, that is, of abstract expressions of the necessary and universal forms of knowledge, can be found not in abstract principles, but only in the immediate consciousness of the forms of representation manifesting itself through statements a priori that are apodictic and in fear of no refutation.[91]

Schopenhauer's principle, however, is not rational love but irrational will, and it is this principle 'in its different aspects' that expresses

> the universal form of all our representations and knowledge. All explanation is a tracing back to this principle, a demonstration in the particular case of the connexion of representations expressed generally through it. It is therefore the principle of all explanation, and hence is not itself capable of explanation; nor is it in need of one, for every explanation presupposes it, and only through it obtains any meaning.[92]

Duration and movement. The second and third tenets of Schenker's philosophy—duration and movement—are the two parameters of human consciousness that music parallels. Hence, duration refers both to subjective time and to musical time. For Schenker, however, the essence of all reality is movement—becoming, eternal flux, change—of which there are two different kinds. First, there is extension, which is movement in the direction of life. Then, there is tension, which is movement in the inverse direction of life. Schenker envisages the first kind of movement as a progressive prolongation of the area of freedom of action, whereas he

---

[91] Schopenhauer, *The World as Will* (note 59), vol. 1, p. 67.

[92] Ibid., vol. 1, p. 73.

regards the second kind of movement as the product of man's freedom that results from his anxiety in being able to make choices. Music, like unfolding life, has extension—horizontalisation, as well as tension—the branching-out or propagation of different musical forms.

Entelechy. The fourth tenet, entelechy, is present in Schenker's notions of will and inner necessity. According to these notions, the function and purpose of the whole organism, in life as in music, determines the overall design or pattern of development. But since Schenker conceives development as creative activity, he believes that this activity is free. The only restriction that arises is from matter itself, namely, from the limitations of human physiology or musical logic (laws of counterpoint). Schenker's conception of development, therefore, must be understood in a twofold manner, namely, in relation to life and in relation to music. On the one hand, creativity refers to the conscious and voluntary efforts of the listener to apprehend new musical forms through choice of prolongation techniques. On the other hand, creative activity refers to musical processes, which, Schenker holds, are free, because music makes itself, since the causes of its unfolding are immanent within the system of tonality.

Although creativity is impelled by the vital, natural power of love, once impelled, creative activity is prolonged by the will. Here again, Schenker's conception of the will is to be understood not only as a characteristic of life but also as a characteristic of music. To comprehend the precise meaning of Schenker's term, '*Tonwille*', however, demands widening the historical context and studying a concept fundamental to his theory of music from the vantage point of the mechanist/vitalist controversy in science. Briefly, the concept in question is *Kraft*, which was referred to originally as living force (*lebendige Kraft* or *vis viva*) but now is used to denote kinetic energy or energy of motion. But the concept of energy did not become important in science until after the establishment of the principle of the conservation of energy enunciated by Hermann von Helmholtz in his important essay, *Über die Erhaltung der Kraft* (1847). The principle of the conservation of energy enabled phenomena such as heat, electricity and magnetism, previously regarded as being due to subtle fluids inconvertible to mechanical motion, to be treated mechanistically. By the end of the nineteenth century, therefore, the limits of natural knowledge coincided with the limits of mechanistic interpretations of reality.[93]

A mechanistic interpretation implies that we can speak of an understanding of some thing only if we succeed in reducing its complex phenomena to simple changes of place or position of ultimate elements and establish universally valid causal rules for these changes. Such an implication is far removed not only from Schenker's but also from Goethe's conception of music and nature as living. Both men situated organic processes between freedom and inner necessity, and both men subjected those processes to no absolute necessity but left a certain amount of free play between different possibilities. As the living organism was dissolved into the concept of mechanism during the nineteenth century, no room remained for a view such as Goethe's, until the Weimar edition of his complete works, published between 1887 and 1919, made available his theory at a time when mysticism in its various forms was prominent; and this may account for the influence of Goethe in the works of musical modernists such as Arnold Schoenberg and Anton Webern, as well as the musical anti-modernist, Schenker.[94]

An explanation of Schenker's concept of *Kraft* is beyond the scope of this chapter, but it is noteworthy that both mechanists and vitalists shared a number of positions in common. These included the notions that matter was discernible by its *Kraft* and not by itself and that there were two kinds of *Kraft*—potential and actual. Potential energy (*Spannkraft*) could be defined in terms of tension; actual energy could be defined in terms of extension. The difference between the mechanists and the vitalists resided chiefly in what they regarded as the basis of *Kraft*. The mechanists thought the basis was matter, whereas the vitalists supposed it was some directed entelechy imposed upon matter. According to Jan Christiaan Smuts, the two positions could be reconciled by means of a redefinition of matter:

---

[93] See, e.g., J. B. Stallo, *The Concepts and Theories of Modern Physics* (London, 1882).

[94] See, e.g., A. Webern, *Der Weg zur neuen Musik* ed. by W. Reich (Vienna, 1960).

> Those who have called the universe creative have implicitly referred to the activity of life and mind in creating new arrangements, meanings and values. It has not been suggested that the physical universe is also creative. The principles of the conservation of matter and energy have effectively barred any such idea. Novelty, originativeness and creativeness are quite inconsistent with the ordinary point of view and the popular ideas of matter as well as the more rigid mechanistic conceptions of science. However, ... in its evolution or creation of the forms, structures and types which characterise it from beginning to end, matter or the physical element in the universe is in a sense as truly creative as is organism or mind. The 'values' of matter or the physical universe arise purely from these structures and forms ... In a very real sense the idea of value applies as truly and effectively in the domain of the physical as in that of the biological or the psychical. In both cases value is a quality of the forms and combinations which are brought about. Whether they are structures resulting from the activities of matter, or works of art or genius resulting from the activities of mind, makes no real difference to the application of the ideas of creativeness and value in either case. Once we get rid of the notion of the world as consisting of dead matter, into which activity has been introduced from some external or alien source; once we come to look upon matter not only as active, but self-active, as active with its own activities, as indeed nothing else but Action, our whole conception of the physical order is revolutionised, and the great barriers between the physical and the organic begin to shrink and to shrivel.[95]

Intuition. The fifth and last tenet of Schenker's philosophy has been the basis for considerable misunderstanding, for it relates to Schenker's scientific method, a method some commentators tend to describe as empiricist. Although Schenker recognises the contributions of music theory drawn from empirical evidence, he frequently exhibits distaste for the scientific procedures employed in the music-theoretic tradition deriving from the late seventeenth and early eighteenth centuries. These procedures relied chiefly on materialistic atomism and on mechanistic interpretations of musical

---

[95] J. C. Smuts, *Holism and Evolution* (London, 1927), pp. 56–57. According to the *OED* (supplement), Smuts coined the term 'holism' to designate nature's tendency to produce wholes from the ordered grouping of unit structures.

reality.[96] In place of mechanistic hypotheses Schenker attempts to introduce organic conceptions and to prove that the vital, guided by love, is in the direction of the voluntary. In proposing a method of grasping this kind of musical reality, Schenker insists that the highest level of musical knowledge is holistic knowledge—the ability to apprehend in one comprehensive perception the inner-relatedness of parts in a musical whole, for he asserts that if 'music exists as an organic creation, we should be able to perceive it'.[97]

Schenker's practical method, which is none other than that described by Plato in the *Republic* and developed afterwards by many different writers, requires two different kinds of 'traversals' in thought.[98] The first traversal is that which leads to the discovery of truth, namely, the eternal forms or archetypal ideas. The second traversal is that which enables presentation of the truth. Although Schenker never drew up a systematic account of his method of discovery, some aspects of it may be gleaned from comments scattered throughout his writings. These comments suggest that Schenker followed the phenomenological method of Goethe, a method that was epitomised by Goethe himself,[99] as shown in *Figure 9.3*.

Holistic knowledge, however, cannot be gained merely through phenomenological enquiry alone, for to achieve a higher level of understanding, it is necessary to find a means for presenting our

---

[96] See, e.g., Schenker/Oster, pp. xxi, 7, 9, 17, 26–27, 106, 112, 131, 132, 136, 138–139. Schenker's barbs are reserved chiefly for the school of Rameau, which advocated what Schenker disparagingly calls 'successive' composition. For his most sustained attack, see 'Rameau oder Beethoven? Erstarrung oder geistiges Leben in der Musik?', *Das Meisterwerk in der Musik: Ein Jahrbuch*...(Münichen, 1930), vol. 3, pp. 9–24; see also Kalib (tr.), *Thirteen Essays* (note 67), vol. 2, pp. 491–518. Schenker/Oster, p. 131, specifically states that his 'theory replaces all these [other theories] with specific concepts of form which, from the outset, are based upon the content of the whole and of the individual parts'.

[97] Schenker/Oster, p. 106.

[98] See, e.g., C. Turbayne, *The Myth of Metaphor* (New Haven, 1962), pp. 29–33 *et passim*.

[99] The fragment is translated in Heinemann, 'Goethe's phenomenological method' (note 63), p. 74.

Phenomenon	
	test empirical
experience	
	test theoretical
law	
	test transcendental
cause	

Figure 9.3. Goethe's phenomenological method

discoveries. To do this, we need adequate ideas, which rely on reason, not perception; that is, adequate ideas deal with logically related, clear propositions that correspond with the abstract generality of mathematics and physics. Hence, a second traversal in thought is required, and the model for this traversal is found in Schenker's treatise, *Free Composition*. In this work Schenker provides adequate ideas, for he presents a deductive system in which all compositions instancing tonality are derived from one of three axioms or *Ursätze* by the successive application of a small number of rules of inference or prolongation techniques.

Only after these two traversals in thought have been made—the one ascending from phenomena to causes, the other descending from causes to phenomena—can we hope to reach the highest level of understanding, namely, holistic knowledge that relies on intuition. Such knowledge is possible because of the archetypal idea—the *Ursatz*, an idea that presents only 'the strictly logical precision' in the coherence between tone successions, regardless of whether the successions are from simple to complex or from complex to simple. For Schenker, therefore,

> It is an inevitable principle that all complexity and diversity arise from a single simple element rooted in the consciousness of the intuition ... Thus, a simple element lies at the back of every foreground. The secret of balance in music ultimately lies in the constant awareness of the transformation levels and the motion from foreground to background or the reverse. This awareness

> accompanies the composer constantly; without it, every foreground would degenerate into chaos... .[100]

Schenker's concept of the *Ursatz* does not determine the chronology of creation or of creative listening, since, according to Schenker, creation 'may have its origin anywhere, in any suitable voice-leading level or tone-succession; the seed, by the grace of God, remains inaccessible even to metaphysics'.[101] To listen creatively, then, we must strive toward holistic knowledge by relying on a number of different procedures, including graphing, carried out as a dialogue or collaboration between ourselves and the music we hear. In this way we may learn to grasp the unity of past, present and future through apprehending in one comprehensive perception the background, middleground and foreground of compositions instancing tonality. A commitment to these three concepts, according to Schenker, enables us to exclude 'all arbitrary personal interpretations' from creative listening.[102]

APPENDIX

Schenker's graphic means of representing the logical relationships between simple tone-successions and more complex ones is illustrated below with reference to a modern score of J. S. Bach's chorale, 'Ich bin's, ich sollte büssen', from *St. Matthew Passion* (BWV 244), shown in *Figure 9.4*. *Figure 9.5* reproduces Schenker's graph of the stages leading to the *Urlinie-Tafeln*, and *Figure 9.6* reproduces the *Urlinie-Tafeln*. Both examples are from Schenker's *Five Graphic Music Analyses (Fünf Urlinie-Tafeln)* with a new introduction and glossary by F. Salzer (New York, 1969), pp. 32–33.[103]

---

[100] Schenker/Oster, p. 18 (Schenker/Jonas, p. 49).

[101] Ibid.

[102] Ibid., p. xxiii.

[103] Michael Kassler has made the emendation in *Figure 9.5*, 1 *Schicht*, second dyad.

Figure 9.4

Key to Figures 9.5 and 9.6

abw. = *abwärts* = descending
aufw. = *aufwärts* = ascending
Ausf. = *Ausfaltung* = unfolding of intervals (i.e., horizontalisation)
Dezimensatz = motion in tenths
Dg. = *Durchgang* = passing motion
Fermaten = pause
Kopp. = *Koppelung* = coupling (which implies transfer of register)
Nbn. = *Nebennote* = neighbour note
Schicht = stage, level
Takte = measures or bars
Tl. = *Teiler* = divider or dividing V, a term used to identify the dominant that precedes an 'interruption' (*Unterbrechung*) indicated in the graph by two upright parallel lines; also employed to indicate V-chords which prolong an underlying tonic
Url Tafel = *Urlinie Tafel* = comprehensive foreground graph which includes the entire analysis in graphic notation of background, middleground and foreground
Ursatz = primitive composition (translated by Oster and others as 'fundamental structure')
Zug = linear progression, but when accompanied by a number it denotes linear progression through a third, fourth, fifth, etc.

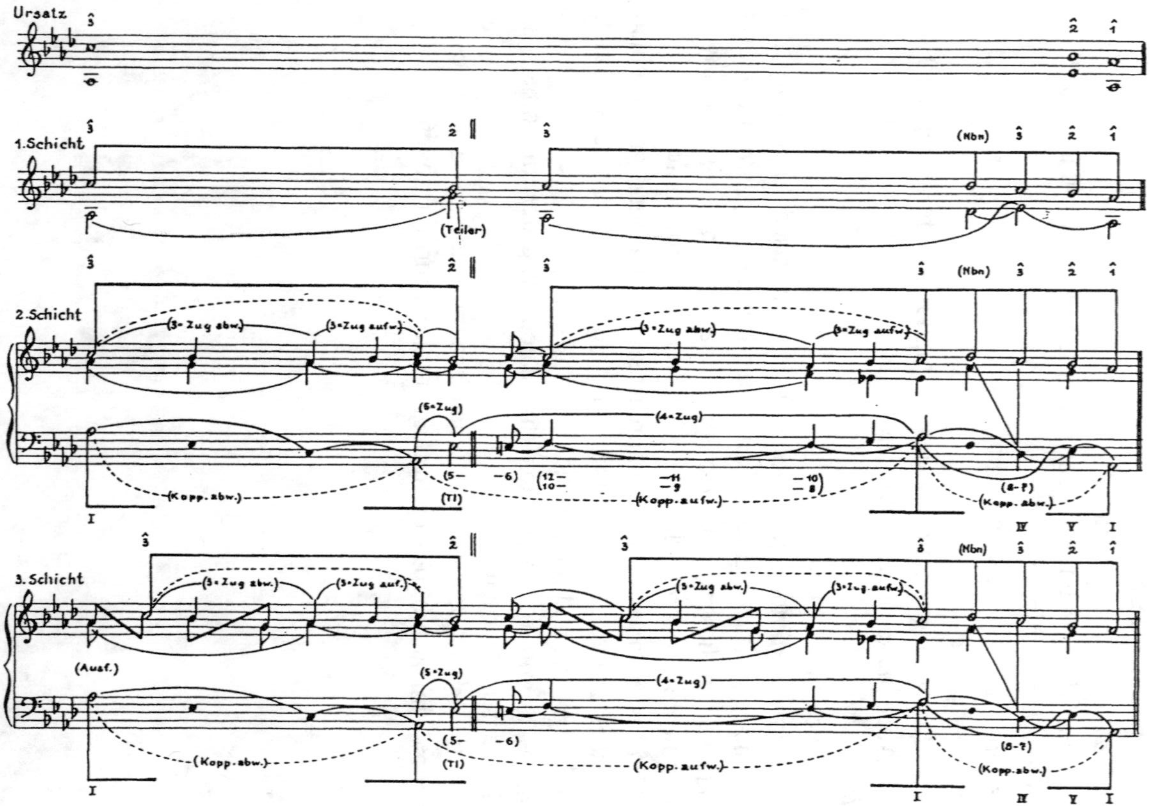
Ursatz
1.Schicht
(Teiler)
(Nbn)
2.Schicht
(3-Zug abw.)
(3-Zug aufw.)
(5-Zug)
(4-Zug)
(Kopp.abw.)
(Kopp.aufw.)
(T1)
(8-7)
3.Schicht
(Ausf.)

Figure 9.5

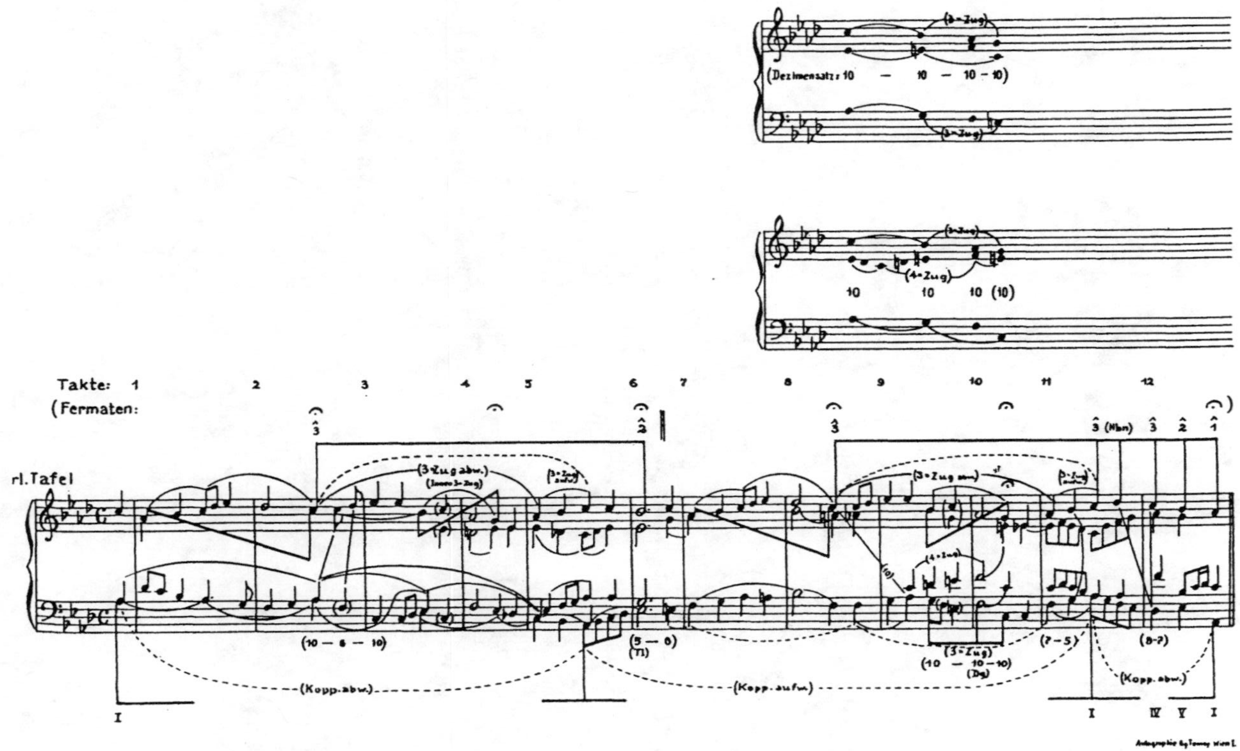

(3-Zug)
(Dezimensatz: 10 — 10 — 10 - 10)
(3-Zug)
(3-Zug)
(4-Zug)
10 10 10 (10)
Takte: 1 2 3 4 5 6 7 8 9 10 11 12
(Fermaten:
rl. Tafel
(3-Zug abw.)
(3-Zug aufw.)
(3-Zug abw.)
(4-Zug)
(10 — 6 — 10)
(5 — 0)
(3-Zug)
(10 — 10 — 10)
(Dg)
(7 - 5)
(8-7)
(Kopp. abw.)
(Kopp. aufw.)
(Kopp. abw.)
I
I IV V I

Figure 9.6

10. THE PROBLEM OF 'MUSICAL EAR'

... the Ear, *or Auditive Faculty ... when it hears ever so well all the particular Sounds ... of a Tune, it does not for all that know ... a Tune to be a Tune.*
Z. Mayne, *Two Dissertations concerning Sense ...* (1728)

INTRODUCTION

Children are exposed to musical sounds at an early age. Indeed, cochlear function is demonstrable as early as the fifth foetal month, when both middle and inner ear structures have reached full adult size.[1] At birth, the neonate prefers to hear its mother's voice, whose vocal characteristics have been learned *in utero*; and by six months, the baby has gained 'a sense of fun and appreciation of musical communication' long before it can speak.[2] Of course, exposure to music differs within families and social groups, as well as from culture to culture. We might ask, therefore:

Does the environment provide sufficient input for children to acquire music?

Is general intelligence plus input from adult example and instruction sufficient to enable children to acquire music?

---

[1] See R. B. Eisenberg, *Auditory Competence in Early Life: The Roots of Communicative Behavior* (Baltimore, 1976); see also J.-P. Lecanuet, 'Prenatal auditory experience', *Musical Beginnings: Origin and Development of Musical Competence* ed. by I. Deliège and J. Sloboda (Oxford, 1996), pp. 3–34.

[2] C. Trevarthen, 'Mind in infancy', *The Oxford Companion to the Mind* ed. by R. L. Gregory with the assistance of O. L. Zangwill (Oxford, 1987), pp. 362–368, p. 364.

Were the principles of the domain—music—learned first and the skills needed to apply them later?

What does it mean to say that 'children know the principles, whether before or after, innately or as the result of learning'?

These questions pertain to what commonly is known as 'musical ear', that is to say, musical *competence*, defined as tacit knowledge and skill. But they also relate to musical *creativity*, in part because theorising about creativity has been dominated by a concern with *sources*, rather than with *procedures* that lead to a product which, contemporaneously or later on, is considered to meet conventionally-established aesthetic criteria of excellence. How, then, are we to approach the problem of musical ear?

I shall answer this question in a preliminary way by quoting, then discussing briefly, a spoken introduction to a twelve-year-old composer, transcribed from the first track of a private recording of some of the child's music. Since the young composer, now grown up, prefers to remain anonymous, let us call him X. The speaker is the noted American composer, Henry Cowell.[3] Here is most of what Cowell has to say:

> [X is] primarily interested in musical composition, not in other aspects of music, except as they pertain to it. And he acts very much like my own colleagues in their attitudes toward things. For instance, music to some people means expert performance; and composers like to hear other people play their works perfectly, but they themselves very seldom are willing to put the energy into practice which will cause ... [them to become] expert virtuosi. They like to be able to play music in order that you may hear it so that they can show the music but not so that you will listen to them play. I think this is X's attitude. He plays very reasonably, and he plays so that you will hear the music. He doesn't play so that you'll hear him playing. And, again, it is worthwhile to point out that he has been at this pursuit for a number of years very steadily. He is therefore—I present him as—a composer of experience.

[3] H. Cowell, 'Introducing X' [my title], Concert of X's music, recorded privately, New School for Social Research, New York, 12 January 1953.

> I have known of these experiences now for sometime myself, because we occasionally work together. And ever since I have known him, he has always composed virtually everyday, rain or shine, ... school day or vacation day, the only protracted exception to this being when he went to the music camp last summer. He found that you can't compose in music camps—they make no provision for this. Obviously, he was not permitted outside the camp; and within the confines of the camp, there was always somebody practising a trumpet or some other ... instrument, so that you couldn't possibly compose.
>
> I would also like to point out that he does not improvise his compositions but that having positive pitch and being able to hear music mentally ... he composes as composers are supposed to, by listening to the sounds in mind and transcribing them on paper and then perhaps trying them over just in order to make sure that there hasn't been a mistake somewhere or just on account of general interest. But, in any case, he is a composer rather than an improviser. He is a person who has stuck to this steadily over a number of years, and he is interested in the elements of musical composition.
>
> He [also] is interested in different musical forms. He has experimented with writing in old musical styles, intermediate musical styles and modern musical styles. And these things have proven very beneficial. Just about the time when I, as a teacher, thought that maybe it would be of interest if he learned a little more about classical music, he would come and say ..., 'I think it would be good for my musical progress if I were to imitate the style of Haydn', ... or something like this, so that one would never have to say anything ... but just ... see that this style was written ... correct[ly] in the part-writing and so on, so that elements [that] are to be found in the music of Haydn really were present. And he has ... not favoured any one type of music over others but has interested himself in all sorts of musical ingredients from Pachelbel to Kabalevsky and with a great many other composers in between.

In his informative introduction to the creativity and competence of X, Cowell distinguished between two types of creative process—that of the improviser and that of the composer. But it would be an error to interpret his distinction as a kind of dualism, in which improvisation would be conceived as a spontaneous, unpremeditated, unwritten-down process and composition, as a memorised, premeditated, written-down process. Indeed, this kind of dualism has no empirical basis, as is demonstrated again and again by studies of western and non-western improvisation in both so-called 'art' music and so-called 'popular' music (e.g., jazz).[4] According to these studies, creativity as improvisation involves constant practice in using well-defined models, for example, thematic or rhythmic formulae, patterns, schemes, etc.; it also involves conventional rules for realising the performance, even if these rules are not verbally articulated or codified in writing.

In the case of X, creativity as composition also involved constant practice in using well-defined models, for example, old, intermediate and modern musical styles; it also involved tacit rules for detecting mistakes in part-writing as well as in realising the mentally-composed work on paper. In societies with a strong utilitarian *ethos*, however, creativity as composition is rare, because music is not valued and encouraged in and for itself. Indeed, in such societies few schools offer cumulative, sequential courses in music similar, say, to those in mathematics. Fortunately for X, his parents valued and encouraged musical experiences (his mother was a music teacher).[5]

Being exposed to such experiences, even *in utero*, he early acquired the competence called 'musical ear' and, hence, could 'hear' music in his head. In addition, there was his curiosity—the desire to know how sounds fit together in combinations, patterns and structures and to know why there is or is not a goodness of fit. Constant practice and experimentation, the outward signs of this striving to know, provided a means for increasing tacit knowledge and for improving requisite skills. In short, all these inseparable attributes (separate only in analysis) form essential, though perhaps not sufficient, bases for

---

[4] See, e.g., F. Tirro, 'Constructive elements in jazz improvisation', *Journal of the American Musicological Society* (1974), 27: 285–305. Space prevents mentioning the important work on computer simulations of aspects of creativity both as improvisation and as composition. For some of this work, see *Computer Music Journal*.

[5] Not all precocious children have had supportive parents and friends. See, e.g., P. Olleson, 'The Wesleys at home: Charles Wesley and his children', *Methodist History* (1998), 36: 139–152, for the suspicions regarding music and musicians that formed the climate of opinion during the childhood of the composer, Samuel Wesley.

the creativity that produces new and sometimes excellent musical works.

You will note that, in my approach to the problem of musical ear, I have had no recourse to one or the other of two traditional theories of the sources of knowledge. These two theories are rationalism, which conceives mind as innately inscribed with knowledge, and empiricism, which conceives mind as a *tabula rasa* written on by experience. In contemporary cognitive science the nativist theory is held by those for whom the brain is a serial, von Neumann, machine: mind (program) is symbolic, and data is processed top down; whereas the non-nativist theory is held by those for whom the brain is a parallel processor: mind is non-symbolic, and data is processed bottom up.[6] One cannot claim, therefore, that the distinction between rationalism and empiricism is dated. Indeed, to paraphrase the words of the neuroscientist, Steven Rose, cognitive philosophers and psychologists are prisoners of their metaphors, and strong hermeneutic traditions are concealed within their scientific practices.[7] There are, of course, variants of the two main theories, and for this reason the twofold distinction may be criticised as inadequate. Nevertheless, it is useful as a starting point for exploring further the

---

[6] These two positions are exemplified in the work of Jerry Fodor, on the one hand, and in that of the so-called 'connectionists', on the other. For some aspects of the debate, see J. Fodor, 'Do we have it in us? The failure of new attempts to explain how the cognitive brain works', *Times Literary Supplement* (May 16, 1997), pp. 3–4. To develop his claim that the mind is multi-stage, P. Johnson-Laird, *The Computer and the Mind: An Introduction to Cognitive Science* (2d edn., London, 1988), modelled the brain on the architecture of both serial and parallel processors. In ontogeny (history of the individual) conscious mind is structured top down in a neo-Lamarckian fashion, whereby frequent experience (e.g., skill acquisition) becomes habitual and automatic. But in phylogeny (history of the species) unconscious mind is structured bottom up in a neo-Darwinian fashion, whereby creativity evolves from the arbitrary combination of elements followed by constraints that filter out products that are unviable. Although this theory may seem to offer a compromise between rationalism and empiricism, it actually is a type of empiricism (for the type, see the section on empiricism below).

[7] Rose is a prisoner of his own metaphors—in particular, his conception of the brain as tape recorder and memory as a magnetic tape. See S. Rose, *The Making of Memory: From Molecules to Mind* (London, 1993), pp. 203, 205, 319, 322–330.

problem of musical ear. I shall therefore retain the traditional distinction though improve it by identifying some variants, so as to show how the questions posed at the outset of this essay have been answered.

The Principles-before Model: Varieties of Rationalism

Rationalist theories of knowledge and learnability are indebted to Plato's doctrine of reminiscence, according to which learning something is merely recognition of what one already knew innately, so that truth, which occurs by the coincidence of knowledge shared innately by humans, is manifested by a process of instantiation or exemplification in particulars—most especially, in naming and in rational discourse that are in harmony with reality.[8] Plato, however, may be regarded as a strong rationalist, since he held that knowledge is innate, and experience merely 'triggers' recognition. In the eighteenth century, Kant gave rationalism a weaker interpretation by admitting a role for experience and then by indicating how knowledge derived from that experience fits into the pattern of innate knowledge.[9] According to him, the way in which the human mind comes to be conscious of the world is inescapably determined by a structuring framework that conditions how the world really appears to us but which cannot show us how reality might appear in itself. He thus introduced the idea that perspective is inescapable, since we have no way of knowing reality as it really is, independent of our structuring framework that we bring to experience.

Despite these differences, both strong and weak rationalists hold that perception is an act of intellectual apprehending or cognising, because the mind is natively disposed to transfer its attention from imaginable objects to unimaginable concepts—concepts which do not lend themselves to using some kind of quasi-optical inner sense that visualises, pictures, imagines or introspects.

---

[8] G. Matthews, *Plato's Epistemology and Related Logical Problems* (London, 1972), reproduces and discusses some of the relevant passages from Plato's *Meno* and *Phaedo*.

[9] I. Kant, *Critique of Pure Reason* [1787] tr. by N. K. Smith (London, 1958).

Rather, it is the organism itself that acts as a mediator between reality and the mind. For example, two strong rationalists, Philip Barford and Victor Zuckerkandl, offered an account of music as an abstract Platonic Idea that a child discovers in the same way that a mathematician discovers the properties of numbers—by intuition.[10] On their view, music is connected neither with the external world nor with the internal world of feeling. Rather, objects as well as persons possess something intrinsic—for Barford, an 'underlying structure', for Zuckerkandl, an 'inner life'. Musical thought is thus a discovery of knowledge within oneself 'in precisely that way which is symbolically subsumed in the primary relationships of geometry'.[11]

Similarly, but less strongly, the French psychoanalyst, André Michel, claimed that musical thought, like conceptual thought, 'is a grasp of consciousness, inferred from unconscious realities'.[12] Like other depth psychologists, therefore, Michel assumed that the mind is stratified and that reality—innate knowledge—is found at the deepest level.[13] For him, the task of the psychoanalyst is to 'penetrate deeply' into the human underground to find the 'organ' or 'root' that feeds 'all branches of music, makes them akin and unifies them'.[14] *Natural* musicality—innate musical knowledge—grows from that 'root', whereas *cultural* musicality—acquired musical knowledge—derives from the exercise and development of auditory functions and tonal aesthetic values which furnish, later on, the material and structure of adult consciousness. But if there is no development of cultural

---

[10] P. T. Barford, 'The concept of Bach', *The Music Review* (1962), 23: 261–270; and V. Zuckerkandl, *Man the Musician: Sound and Symbol* (Princeton, 1973).

[11] Barford, ibid., p. 267.

[12] A. Michel, 'Psychoanalysis of music', *The Music Review* (1950), 11: 253–276, pp. 262–263.

[13] Although his depth psychology is eclectic, Michel, ibid., pp. 253–261, utilised Freudian psychoanalytic theory to propound a theory of musical 'style', i.e., personality, based on different stages of development—oral (e.g., Chopin), anal (e.g., Ravel), phallic (e.g., Liszt) and latent (e.g., teachers 'who ignore the personal experience of composing'). This reductionist approach entails judgments such as the following: 'The musical substance … of Liszt proceeds … by successive erections.'

[14] Ibid., p. 256.

musicality, the universal natural 'root' withers away from lack of exercise.

The above examples illustrate an important feature of all nativist theories: because constraints are conceived as the knowledge a learner brings to the learning task, the problematic term is 'knowledge'. Nativists agree that what is brought to the task of acquisition must be domain-specific knowledge, that is, knowledge which is particular to a specific domain like music or mathematics. But they lack agreement as to what constitutes *the particular content* of domain-specific knowledge. This lack of agreement enables us to distinguish three varieties of nativism: concept nativism, principles nativism and abilities nativism.

Concept nativists believe that concepts from a domain are innate. Hence, they must give a precise account of the notion of ideational or conceptual content. For example, does possessing a concept require having a word or other symbol for it? Can there be a single isolated concept or must there be a whole network? What are the criteria for having a musical concept? Answers to some of these questions may be found in the work of the experimental psychologist, Geza Révész, who defined musicality as 'the ability to distinguish, apprehend, and recognize acoustic-musical relationships'.[15] Although this ability unfolds through experience and practice, it depends upon an innate sense of pitch discrimination that is prior to, and a prerequisite for, musicality.

Révész, however, identified two types of pitch discrimination. On the one hand, there is 'a general psychological faculty', possessed by all healthy individuals, whereby pitches are recognised and identified according to their 'highness' and 'lowness' in the musical scale. On the other hand, there is 'a special natural gift'—a 'musical ear', possessed by comparatively few persons (mostly men, according to this author), whereby pitches are recognised and identified directly, without explicit reference to, or comparison with, other pitches. Both types of pitch discrimination, general and special, are part of the

[15] G. Révész, *Introduction to the Psychology of Music* tr. [from the German of 1944] by G. I. C. de Courcy (Norman, Oklahoma, 1954), pp. 95–111; see also pp. 35–36, where he dismisses non-nativist theories that attempt to explain the psychophysical connection, possibly under the influence of his teacher, Carl Stumpf, also a nativist.

innate memory store. In the case of the general faculty, the concepts, high and low, are innate. But in the case of the special faculty, what is innate is the concept of a 'definite aural standard', namely, the international standard pitch of 440 Hertz, so that the ability to detect readily slight deviations from standard pitch is learned. On this account, 'musical ear' is the innate faculty of absolute pitch.

What, then, are the criteria for assessing whether or not a person has this kind of 'musical ear'? According to Révész, statistical studies show that absolute pitch tends to be inherited; that it is evident in children by the beginning of the third year, who demonstrate pitch-naming ability as soon as an appropriate situation arises; and that this pitch-naming ability is not equally developed in all who have it (differences being noted in respect to the extensity of the tone domain and in respect of timbre). In short, naming is the behaviour (performance) that manifests the innate faculty (competence) of absolute pitch.

Although the term 'absolute pitch' is now common coin, there has been little historical study of the development of this concept.[16] One might conjecture, therefore, that the concept is an artefact that emerged in the 1840s with the establishment of absolute standards of pitch—standards that furnished the then new experimental psychology with specifications for tonal stimuli.[17] Central to the new approach was the belief that frequencies of vibration that give pitches are strictly periodic. With the advent of electro-acoustics, however, it was found that semantic or aesthetic information transmitted by

---

[16] For historical studies of absolute pitch, see A. G. Pikler, 'History of experiments on the musical interval sense', *Journal of Music Theory* (1966), 10: 54–95. For an analysis of the concept, see L. K. Miller, *Musical Savants: Exceptional Skill in the Mentally Retarded* (Hillsdale, New Jersey, 1989), pp. 41–63. The need for further investigations in this area is urgent, because so many theories of musical ear focus on the pitch sense. The topic, however, is highly complicated; for it is not always easy to discover the hermeneutic traditions of those who explain pitch sense by recourse to neuro-mythological structures such as a series of brain circuits tuned to each frequency or of those who argue that there is an internal scale of relative pitch conceived as a movable conceptual grid or template that specifies pitch relations among notes of a culture's scale.

[17] See E. G. Boring, *Sensation and Perception in the History of Experimental Psychology* (New York, 1949), p. 320.

certain kinds of vibrations is not periodic, so that the tiny deviations from periodicity are better analysed by statistical rather than classical mathematics. Thus, as Fritz Winckel pointed out, the 'standpoint of the absolute' was to be relinquished.[18]

W. Dixon Ward and Edward M. Burns provide support for the above conjecture, for they point out that the 'ability' called 'absolute pitch' (AP) 'has been subjected to scientific scrutiny for about a century', especially by the tone psychologist, Carl Stumpf.[19] The same authors claim that 'we now know quite a bit about the characteristics of AP', even though there still is 'an enigma' as to 'the genesis of the ability':

> why do some people identify pitches with no apparent effort while others must engage in strenuous training to develop AP? Indeed, there was doubt that it was even possible to 'learn' AP until Paul Brady did so [as an adult] in 1970. For this reason, its possession was for many years regarded with pride, as if it indicated that its owners were mysteriously gifted.[20]

Ward and Burns favour a learning theory without denying there might be a genetic component. They also believe that the nativist theory of AP died out in 1957, thus ignoring, for example, not only the beliefs of some cognitive scientists but also recent endeavours at gene mapping that led some researchers to speculate there might be a gene for absolute pitch.[21]

---

[18] F. Winckel, *Music, Sound and Sensation: A Modern Exposition* tr. by T. Binkley (New York, 1967), p. 164. See also J. C. Kassler, 'Musicology and the problem of sonic abuse', *Music, Sensation and Sensuality: A Collection of Essays* ed. by L. P. Austern (New York, forthcoming). Cowell's term, 'positive pitch', does not imply an absolute standpoint.

[19] W. D. Ward and E. M. Burns, 'Absolute pitch', *The Psychology of Music* ed. by D. Deutsch (New York, 1982). Their reference is to C. Stumpf, *Tonpsychologie* (Leipzig, 1883).

[20] Ibid., p. 434. Paul Brady was author of 'Fixed-scale mechanism of absolute pitch', *Journal of the Acoustical Society of America* (1970), 48: 883–887.

[21] See, e.g., L. Slattery, 'Minds to reckon with just don't add up', *The Australian* (March 22, 1999), p. 5, who reports on the claims of a scientist at the Australian National University who assumes that 'perfect pitch', along with perspectival drawing and lightning mathematical calculations, are accessed from 'lower

A second variety of nativism postulates that principles, not concepts, are innate and underlie children's ability in a specific domain. In the domain of music these principles are taken to serve two purposes. On the one hand, the principles enable a child to recognise and distinguish instances of music from other happenings in the environment; on the other hand, they guide a child's own musical behaviour by serving as a standard for correct musicality. What, then, are the principles? The antiquary, Daines Barrington, identified two principles that determine a natural 'genius' or 'disposition' for music.[22] First, there must be a sense of pitch defined as recognition of the keynote either *(a)* from hearing the first two or three bars of a composition, or *(b)* from knowing what must be the concluding bass note, or *(c)* from retaining in memory every successive note of the three bars and from thence determining the key. Second, there must be a sense of time defined as recognition of the commencement of a musical bar without beating or counting.

For Barrington, then, the term 'musical ear' denotes the innate principles of 'tune' and 'time', since he supposed that these principles are the origin, as well as the guide of premature or untaught 'talents' (musical behaviours). In the nineteenth century craniologists adopted a similar approach, because they assumed that 'faculties'—including those of tune and time—were innate and localised in specific 'organs' on the cortex of the brain. On these premisses one craniologist, Johnann Caspar Spurzheim, could argue that the ear was not the organ of musical perception, because, just as singing birds 'hatched by strange females, sing naturally, and without any instruction, the

levels of "raw" information', that they are related to 'unconscious problem solving' and that 'savants' (i.e., autistic and other brain damaged children) have 'privileged' access to the lower levels. See also, e.g., O. Craig and S. Connor, 'Born to succeed', *The Australian* (August 14, 1997), p. 10, who report on the study of scientists at the University of California who seek a single gene for 'perfect pitch' (the study was based on sixty families in which absolute pitch seems to have been 'handed down through the generations').

[22] D. Barrington, 'Account of a very remarkable musician', *Philosophical Transactions of the Royal Society* (1770), 60: 54–64.

song of their species as soon as their internal organization is active', so too musicians who have lost their hearing continue to compose.[23]

Principles nativism also underpins the work of the musicologist, John Blacking, who, from the 1970s, argued that music is part of a human bio-grammar that includes language and dance.[24] He conceived this bio-grammar as a universal set of competences, 'deep' structural principles, from which is generated varying surface forms in different societies and historical moments, as well as among individuals within a given society. We might suppose, therefore, that Blacking was indebted to the linguist, Noam Chomsky, who argued that native speakers have a tacit knowledge (competence) of the principles underlying well-formed sentences and that these principles are capable of generating a potentially infinite variety of human languages.[25] Assuming that there is a genetically determined language 'faculty' as one and only one component of mind, Chomsky argued that a principle of language is universal if we are led to postulate it as a 'precondition' for the acquisition of a single language. For him, that precondition is universal grammar, a system that specifies the phonetic, syntactic and semantic properties of an infinite class of potential sentences. Universal grammar is thus a representation of a child's intrinsic competence; but in the acquisition of a specific (culturally relative) language, a child also develops 'performance systems' for putting this innate knowledge to use.

In a third variety of nativism, abilities from a domain are presumed innate. But, as we shall see in the next section, arguments of empiricists also appeal to abilities as aids in the acquisition of musical knowledge. Hence, since non-nativist theories, implicitly at least, do not deny that humans come equipped with innate abilities, a distinctively nativist position depends on being able to give substance to the idea that the precursors underlying domain-specific knowledge

---

[23] J. Spurzheim, *The Physiognomical System of Drs. Gall and Spurzheim: founded on an anatomical and physiological Examination of the Nervous System in General, and of the Brain in particular; and indicating the Dispositions and Manifestations of the Mind*...(London, 1815), pp. 104–105.

[24] J. Blacking, *Music, Culture and Experience: Selected Papers*...ed. by R. Byron (Chicago, 1995).

[25] See, e.g., N. Chomsky, *Aspects of the Theory of Syntax* (Cambridge, 1965).

neither reflect nor are derived from seemingly related general abilities and predispositions.

In attempting to satisfy this condition, the music historian, Charles Burney, asserted that 'premature' or 'supernatural' disposition towards music originates from two innate abilities—an acute ear and an infallible memory.[26] He claimed that these natural abilities are evidenced by a child's musical behaviour—specifically, by ready detection of wrong notes, and singing or playing 'by ear' either *(a)* the keynote of a tune, or *(b)* the first three notes of a tune or *(c)* an entire tune. Lullabies heard in the nursery neither stimulate the attention of children to melody nor accelerate the display of 'musical talents' at an early age, because art is governed by laws built on productions and effects 'as the most polished part of mankind have long agreed to call excellent' and to which only small approaches towards perfection can be made by infants 'in a state of nature'. Without the aid of reason and perseverance, an infant can depend for his guides only on acuteness of ear and infallible memory; but these natural abilities require cultivation, else they may be exhausted or enfeebled, 'their energy and vigour' impeded by 'self-complacence', by 'an injudicious course of study', 'by ill-chosen models' or by 'want of counsel'.

In the nineteenth century Darwin's cousin, Francis Galton, expounded a more extreme version of abilities nativism in his study of 'genius', which he supposed was inherited chiefly from the male line.[27] His graded continuum of 'natural abilities', ranging from stupidity at one end to genius at the other, was based on a belief that natural ability denoted instincts—dispositions or inclinations, as well as intellect—mental faculties or powers, since the two aspects, together,

---

[26] C. Burney, 'Account of an infant musician', *Philosophical Transactions of the Royal Society* (1779), 69: 183–206.

[27] F. Galton, *Hereditary Genius: An Inquiry into its Laws and Consequences* (London, 1869), who, like Darwin, relied on pangenesis for his theory of heredity. Later, pangenesis was replaced by Mendelism; see, e.g., C. C. Hurst, 'Mendel's law of heredity and its application to man', *Transactions of the Leicester Literary and Philosophical Society* (1908), 12: 35–48, and 'Mendelian heredity in man', *The Eugenics Review* (1912–13), 4: 1–25, who, in the second article, altered 'musical sense' to 'musical temperament'.

> urge and qualify a man to perform acts that lead to reputation. I do not mean capacity without zeal, nor zeal without capacity, nor even a combination of both of them without an adequate power [i.e., energy] of doing a great deal of very laborious work. But I mean a nature which, when left to itself, will, urged *by an inherent stimulus*, climb the path which leads to eminence and has strength to reach the summit—one which, if hindered or thwarted, will fret and strive until the hindrance is overcome, and it is again free to follow its labour-loving instinct.[28]

Although Galton allowed that some few persons possessed 'universal gifts', most persons of genius possessed 'special gifts' that were domain-specific. He noted, for example, that musicality manifested itself earlier than gifts in other domains, since in addition to an inherited instinctive disposition for music, there also is an inherited 'musical faculty', namely, 'musical taste'.[29]

A century later, the neurologist, Norbert Wertheim, distinguished musicality into two components, receptive and expressive. The former alone constitutes musical ear—a 'faculty' that is 'an inborn capacity of the human brain', 'common among human beings' and 'independent of education or culture which can refine but not produce it'.[30] According to Wertheim, the inborn faculty is 'in fact' three 'basic abilities', one general and two special. The general ability, 'probably' of biological origin, is a rhythmic sense, the function of which is 'superior integration'. The other two special abilities are a sense of sounds and a sense of musical intervals. The function of the first is correct perception of the four elementary physical properties of a sound—pitch, loudness, timbre and duration, with absolute pitch being a 'refinement' of this ability, whereas the function of the second is recognition of tonal distance between different sounds in a melody or chord. These three abilities or 'feelings' are necessary prerequisites of musicality, because in the

---

[28] Galton, ibid., pp. 37–38, italics mine.

[29] Ibid., pp. 79, 237.

[30] N. Wertheim, 'Is there an anatomical localisation for musical faculties?', *Music and the Brain: Studies in the Neurology of Music* ed. by M. Critchley and R. A. Henson (London, 1965), pp. 282–297.

absence of any one, 'there can be no musical function', that is to say, no expressive aspect, which Wertheim defined as the 'aptitude to convert musical perception into emotional or intellectual content'. This expressive aspect is 'built up empirically, or through persistent study which permits the elaboration of a special praxic ability based upon the three receptive components'.

THE PRINCIPLES-AFTER MODEL: VARIETIES OF EMPIRICISM

Empiricist doctrines of knowledge and learnability are perhaps indebted to Aristotle's theories of sense perception, memory and logic of science.[31] From these theories it was held that since *nihil est in intellectu quod non prius fuerit in sensu*, knowledge is attained by induction, that is to say, by being induced or led on to recognise a principle as evidently true or a concept as universal by abstraction from what is sensorily given. Aristotle, however, may be considered a narrow empiricist, since he emphasised the metaphysics of observables and conceived induction as conceptual reproduction of what has been seen.[32] In more recent times empiricism has been given a broader interpretation by those who emphasised conceptual or theoretical aspects of science such as hypothesising, theory construction and assessment and the role of concepts in the formalism of a system.[33]

Despite these differences, both narrow and broad empiricists insist that all knowledge is derived from, and relates to, the empirical

[31] Whether Aristotle was responsible for an empiricist theory of knowledge is debatable; see, e.g., D. W. Hamlyn, *Sensation and Perception: A History of the Philosophy of Perception* (Cambridge, Massachusetts, 1961), pp. 17–18.

[32] This is the traditional interpretation of Aristotle, for an early instance of which see Chapter 4 *supra*. For a non-traditional interpretation, see M. Nussbaum, 'Essay 5. The role of phantasia', *Aristotle's De motu animalium*...(Princeton, 1978), pp. 221–269.

[33] See, e.g., W. S. Jevons, *Pure Logic and Other Minor Works* ed. by R. Adamson and H. A. Jevons (London, 1890), pp. 197–299; and K. R. Popper, *Conjectures and Refutations: The Growth of Scientific Knowledge* (London, 1963). See also F. Enriques, *The Historic Development of Logic* tr. by J. Rosenthal (New York, 1929), pp. 235–239 *et passim*.

world, the world of which we have experience. They therefore deny that there is anything that can be called 'innate ideas', meaning by this, substantial propositions—beliefs as opposed to predispositions to behave in certain ways. Hence, their doctrines are not upset by supposing there are preferences, e.g., for certain sound orders that are very general and that might be a feature of human behaviour. Instead, the contrast they insist on is between knowledge, in the sense of propositional knowledge, and features of behaviour, such as the tendency to represent on paper a tune in a certain way that may vary culturally. Nevertheless, empiricists believe that children must be well endowed with general capacities to take advantage of inputs from experience, even though such capacities are not domain-specific.

For empiricist theories, therefore, 'experience' is the problematic term. For example, in demanding that we base knowledge upon experience, empiricists must either define what kind of thing experience is supposed to be, why it is stable and why it serves well as the foundation of knowledge; or they must admit some role for a principle by which experience is selected and ordered and then must indicate how knowledge derives from that principle.[34] From their disagreements about how these tasks are to be accomplished, we are able to distinguish three varieties of empiricism: psychological or empiricism without structure, physical or empiricism determined by physiology and biological or empiricism determined by evolution.

Psychological empiricism holds that external objects are the determinants of knowledge. Its main representative is the philosopher and physician, John Locke, who conceived knowledge as inductive assembly: knowledge arises from experience by the application of step-by-step inductive operations, and abilities are somehow transferred from culture (environment) to initially blank minds in the form of symbol systems such as music or mathematics or language. Such knowledge as we do acquire takes its form from the regularities

---

[34] For an example of the first approach, see H. Cantril, 'Concerning the nature of perception', *Proceedings of the American Philosophical Society* (1960), 104: 467–473, who identifies four varieties of human experience: *(a)* as reflected in on-going naïve experience; *(b)* as described; *(c)* as analysed and conceptualised; and *(d)* as abstracted for specification.

in our experience of the world, so that knowledge grows adventitiously where it happens to fall, and all that is needed by way of mind is a mechanism to receive input and a mechanism to 'associate' or assemble it in a fashion that abstracts its steady features.

Locke described the first mechanism as the 'faculty of laying up, and retaining the Ideas, that are brought into the Mind' from sense perception.[35] Moreover, he noted that this retentive mechanism is common not only in humans but also in several other animals, for,

> to pass by other Instances, Birds learning of Tunes, and the endeavours one may observe in them, to hit the Notes right, put it past doubt with me, that they have Perception, and certain Ideas in their Memories, and use them for Patterns. For it seems to me impossible, that they should endeavour to conform their Voices to Notes (as 'tis plain they do) of which they had no Ideas. For though I should grant Sound may mechanically cause a certain motion of the animal Spirits, in the Brains of those Birds, whilst the Tune is actually playing; and that motion may be continued on to the Muscles of the Wings, and so the Bird mechanically be driven away by certain noises, because this may tend to the Bird's Preservation: yet that can never be supposed a Reason, why it should cause mechanically, either whilst the Tune was playing, much less after it has ceased, such a motion in the Organs of the Bird's Voice, as should conform it to the Notes of a foreign Sound, which imitation can be of no use to the Bird's Preservation. But which is more, it cannot with any appearance of Reason, be supposed (much less proved) that Birds, without Sense and Memory, can approach their Notes, nearer and nearer by degrees, to a Tune play'd yesterday; which if they have no Idea of it in their Memory, is now no-where, nor can be a Pattern for them to imitate, or which any repeated Essays can bring them nearer to. Since there is no reason why the sound of a Pipe should leave traces in their Brains, which not at first, but by their after-endeavours, should produce the like Sounds; and why the Sounds they make themselves, should not make traces which they should follow, as well as those of the Pipes, is impossible to conceive.[36]

---

[35] J. Locke, *An Essay concerning Human Understanding* [4th edn., 1700] ed. by P. H. Nidditch, (Oxford, 1975), pp. 154–155.

[36] Ibid.

In addition to a retentive mechanism, there also is an associative mechanism that determines the way in which ideas, feelings and movements are connected as successions in the mind or as actions of an individual or as (in the above example) the process of establishing such connections.[37] To illustrate this mechanism, Locke had recourse to a musician for whom a particular tune has become second nature. Hence, when that tune

> but once begin in his Head, the *Ideas* of the several Notes of it will follow one another orderly in his Understanding without any care or attention, as regularly as his Fingers move orderly over the Keys of the Organ to play out the Tune he has begun, though his unattentive Thoughts be elsewhere a wandering. Whether the natural cause of these *Ideas*, as well as of that regular Dancing of his Fingers by the Motion of his Animal Spirits, I will not determine, how probable soever by this Instance it appears to be so: But this may help us a little to conceive of Intellectual Habits, and of the tying together of *Ideas*.[38]

Once a mind is stocked with tunes ('Intellectual Habits'), an appropriate stimulus will cause a tune to play automatically as trains of musical ideas in the head or as orderly movements of the fingers on an instrument, even when conscious attention wanders. Hence, 'operations' such as thinking, determining the will and moving the body are nothing but 'Trains of Motion in the Animal Spirits, which once set a going continue on in the same steps they have been used to, which by often treading are worn into a smooth path, and the Motion in it becomes easy and as it were Natural'.[39]

For Locke, then, knowledge is determined by external objects of knowledge: The sense organ (the brain) assimilates the pattern of the object without its matter just as wax assimilates the pattern of the seal without the bronze or gold in which it is cut. Although Locke

---

[37] Locke's principle of associative or inductive assembly was first described in the fourth edition of his *Essay*. Commonly referred to as 'the association of ideas', this principle assumes that complex ideas are formed from simple ones by repeated juxtapositions in experience.

[38] Ibid., p. 269.

[39] Ibid., p. 369.

thought it probable that 'the Constitution of the Body does sometimes influence the Memory', he declined to consider how our constitution or the make of our animal spirits influence memory or 'whether the Temper of the Brain make this difference, that in some it retains the Characters drawn on it like Marble, in others like Free-stone, and in others little better than Sand'.[40] Later on, however, he was less tentative, when he treated the cause of obscurity in simple ideas by an analogy to wax differently tempered.[41] It is well known, of course, that Locke's scepticism about structure led him to emphasise function—in the case of our examples, the functions of memory—retention and association. He thus conceived musical ear psychologically as quick recollection—the recovery of latent habits (patterns of tunes) stored in memory.[42]

Not long after Locke's death, another philosopher and physician, David Hartley, developed an empiricism determined by parallelistic psychophysiology, for he postulated that the physical basis for sensation and memory were 'vibrations' of particles in the nervous system that persisted in the form of smaller 'vibratiuncles'.[43] According to his theory, the functions of mind (i.e., the mechanisms of inductive assembly) depend upon vibrations that pass a single sensation instantaneously along the nerve to the medullary particles of the nerve endings at appropriate locations within the brain, where

---

[40] Ibid., pp. 152–153.

[41] Ibid., pp. 363–364.

[42] Paradoxically, this conception has roots in the nativist tradition, since in Plato's *Meno* an example is given of an untaught slave whose latent knowledge is reawakened. In Locke's day one of the Cambridge Platonists substituted a musician for the slave, for which see J. C. Kassler, *Inner Music: Hobbes, Hooke and North on Internal Character* (London, 1995), pp. 115–117. It is not surprising, therefore, that Locke's treatment of musical ear is indebted to the neo-Platonist, Thomas Willis (see Chapter 6 *supra*), who dealt with that subject in *Cerebri anatome* (1664); see W. Feindel (ed.), *The Anatomy of the Brain and Nerves*...(2 vols., Montreal, 1965), vol. 2, pp. 116–121, 141–145. See also P. Gouk, 'Some English theories of hearing in the seventeenth century before and after Descartes', *The Second Sense*...ed. by C. Burnett et al. (London, 1991), pp. 95–113.

[43] D. Hartley, *Observations on Man, his Frame, his Duty, and his Expectations*...(London, 1749).

the single sensation is realised in consciousness and remains as a 'vestige' of itself for a short time. If the vestige is reinforced by repetition of the sensation, it becomes a simple idea, remaining permanently to help form complex ideas by association. The repeated sensation produces a miniature idea of itself (a 'vibratiuncle') that is intellectual rather than sensible. By the process of association, miniature ideas of any given 'synchronous' or 'successive' sensations lodged contiguously in the brain substance exhibit a peculiar inherent power to recall one another when one of them is excited. Moreover, these miniatures may cluster into compounds, and compounds may be further compounded into abstract ideas that are ultimately reducible to simple ideas of sensation and traceable to gross sense experience.

Although the physiological determinant in Hartley's empiricism has been superseded, he remains influential for having developed the central tenets of association—for him, the moral analogue of Newtonian gravitation.[44] These tenets are as follows:

(1) All cognitive processes are conceived as associative.

(2) Association is treated as a stimulus-response connection: a stimulus instigates or elicits a response; the index for the presence of an association is the eliciting of a response; and the regularity with which the stimulus elicits it (or the probability of its occurrence) is the index of the strength of an association.

(3) The data of recall provides the sole evidence concerning the formation of associations. Since an association is formed at one point of time, the evidence of its presence is obtained subsequently; hence, the coherence of psychological functioning is explained by the mechanism that connects individual experiences or actions according to the order in which they occur.

(4) The stimulus-response sequence provides the model for associative processes as unidirectional. A stimulus-response sequence, which is a sample of action, has a forward direction as an

[44] I.e., Hartley, ibid., p. 6 *et passim*, utilised Newton's method of philosophising.

essential characteristic. (Later, the 'law' of forward neural conduction reinforced this notion of a forward direction, or unidirectionality.)

(5) The principle of unidirectionality asserts that an association between *a* and *b* and between *b* and *a* are distinct events—they are two associations. Hence, any departure from unidirectionality was conceived as a source of disorganisation or maladaptiveness (interference). It was assumed that specific 'chaining' of connections alone provides for orderly cognitive functioning, i.e., that no other cognitive operations secure order.

Associationism, therefore, is a form of causal explanation: ideas, feelings and movements must occur in a specific order if they are to be appropriate.[45]

To illustrate this order, Hartley instanced the transition of voluntary actions into 'secondarily' (acquired) automatic ones by recourse to the example of an aspiring performer learning to finger a musical instrument:

> Suppose a Person who has a perfectly voluntary Command over his Fingers, to begin to learn to play upon the Harpsichord: The first Step is to move his Fingers from Key to Key, with a slow Motion, looking at the Notes, and exerting an express Act of Volition in every Motion. By degrees the Motions cling to one another, and to the Impressions of the Notes, in the Way of Association ..., the Acts of Volition growing less and less express all the Time, till at last they become evanescent and imperceptible. For an expert Performer will play from

---

[45] For a critique see S. E. Asch and S. M. Ebenholtz, 'The principle of associative symmetry', *Proceedings of the American Philosophical Society* (1962), 106: 135–163. It does not follow that the ability to think of ideas, feelings and movements in any other order obscures our knowledge of the order in which they did or should occur. Even so, psychologists in the Gestalt school have criticised the associationist principle of unidirectionality by seeking to show not only that that principle is false but also that associative operations partake of the flexibility that characterises cognitive functioning generally. For example, in thinking (i.e., reasoning) we traverse the available terrain in different and opposed directions (i.e., analysis, synthesis). The cause recalls the effect, but the effect also brings the cause to mind. In working on a music-theoretical problem, the succeeding steps bring us to the goal, but we also work back from the goal to the required steps.

> Notes, or Ideas laid up in the Memory, or from the Connexion of the several complex Parts of the decomplex Motions, some or all; and, at the same time, carry on a quite different Train of Thoughts in his Mind, or even hold a Conversation with another.[46]

Although this example of the principle of conscious decay under habituation is not original with Hartley, it was to become the paradigm case for acquired reflex actions of all kinds, including that of speech. Indeed, so widespread was this paradigm that it became known as 'the clavier theory'.[47] If all complex processes of mind were to be explained by mere association of ideas and if these associations were made possible by virtue of a nervous substrate, the chief characteristic of which was a tendency to take on and retain habits, then we have the basis for a mechanistic explanation of both objective behaviour and subjective experience.

Hartley's variety of empiricism was reinterpreted in biological terms after publication in 1859 of Charles Darwin's *Origin of Species*.[48] According to the new interpretation, knowledge is a function of a purely natural organism. The physical basis of this function is the brain and nervous system, which is the product not only of the experience of the individual but also of the species. The problem then becomes: how did the human species acquire knowledge *de novo*? Initially, two theories were proposed to answer

---

[46] Hartley, *Observations on Man* (note 42), pp. 108–109.

[47] The term, '*Claviertheorie*', was utilised in 1842 by Johann Wilhelm Arnold; see F. Fearing, *Reflex Action: A Study in the History of Physiological Psychology* (New York, 1964), p. 158, and J. C. Kassler, 'Man—A musical instrument: Models of the brain and mental functioning before the computer', *History of Science* (1984), 22: 59–92. It is important to note that, although association is usually thought of as the instrument (often the only instrument) of empiricism, introspectionist and Gestalt psychologies also accepted that a conscious state, by repetition and under the 'law' of habit, is telescoped and reduced until the given process is largely or entirely unconscious (i.e., physiological). The main difference between associationism and introspectionism/Gestaltism, therefore, is not a difference in emphasis but a difference in the object and, hence, of the concept as an aggregate of sensations (associationism) or as a unity (introspectionism/Gestaltism).

[48] C. Darwin, *On the Origin of Species by means of Natural Selection: or the Preservation of Favoured Races in the Struggle for Life* (London, 1859).

this question, both of which permitted inferences about ancestors from embryonic stages of descendants.[49] On the one hand, there was the theory of increasing differentiation that called upon a conservative principle of heredity to preserve the early stages of ontogeny in all members of a group, while evolution proceeds by altering later stages. On the other hand, there was the recapitulation theory, which required an active mechanism that pushes previously adult features into progressively earlier stages of descendant ontogenies. It is noteworthy, therefore, that two metaphors have been used to symbolise one or the other theory. One metaphor, from palaeontology, pictures the nervous system in layers or strata; the other, from botany, pictures the nervous system as branches that grow from a common trunk.

The first metaphor occurs in the work of the nineteenth-century neurologist and physician, John Hughlings Jackson, who held that the brain, in the developing individual as well as in the evolution of the species, is built up by the addition of successive and increasingly complex layers, so that each new stratum covers and integrates the underlying structure.[50] At any evolutionary level of the animal kingdom, the higher integrative centres are less rigidly organised structurally than the lower ones—in modern terms, the neurons of the higher centres have more neurons of the lower differentiation level, whereas lower centres have more rigidly organised (canalised) neuron systems. Each neuron system starts its evolution from lower levels of differentiation with characteristic diffuse net-like connections and tends towards the higher stages with few rigidly determined almost linear connections. Integrative capacities may increase in the first part

---

[49] S. J. Gould, *Ontogeny and Phylogeny* (Cambridge, Massachusetts, 1977), p. 3.

[50] J. H. Jackson, *Selected Writings*...ed. by J. Taylor...(2 vols., New York, 1958). Herbert Spencer developed a similar hypothesis by recourse to the antiphonel, a self-acting keyboard instrument with mechanical 'memory', the latter consisting of a pinned 'planchette' placed on top of an existing keyboard. See H. Spencer, *Principles of Psychology* (3d edn., 2 vols., London, 1890), vol. 1, pp. 567–568. Invented by a French mechanician, Alexander Debain, the antiphonel was patented on 29 August 1846 (Patent No. 11, 359) and exhibited in London at the Great Exhibition of 1851. Spencer, however, seems to have been influenced by Étienne Bonnot de Condillac, whose keyboard model is extricated by Kassler, 'Man—A musical instrument' (note 47), pp. 77–80.

of this evolutionary process, but they are progressively lost in the second. It is for this reason that the higher integrative centres are always those relatively newly acquired during phylogenesis, having therefore many neuron systems of relatively early differentiation stages, while the phylogenetically old centres, besides becoming more and more organised (automatic), lose their integrative capacities.

Like Hartley, Jackson assumed that, physiologically, the basis of all nervous activity is sensori-motor or reflex and consists of an adjustment to the environment; and that, psychologically, consciousness, which is directed to some activity, 'accompanies' or 'arises' during reflex actions in which the highest cerebral centres are engaged. Thus, in considering why 'idiots' were 'not much behind other children' in their 'mere taste for music', Jackson conjectured that music is 'probably one of the most automatic of the higher mental operations; hence the significance of the existence of musical faculty in those who have little higher mental faculty'.[51] And in considering why patients who have a slight fit when employed in playing a well practised tune 'may go on doing that automatic thing—may continue playing correctly whilst unconscious', he observed that the 'automatic action had, so to speak, possession of the mind, and consciousness was not concerned in it before the paroxysm occurred'. He also noted that even a healthy person may 'play a simple well-learned tune when talking of something else', thereby indicating that 'a transitory lack of consciousness might not interfere with his performance.'[52]

Jackson's hierarchy of structures that cannot be conceived separately leads to the principle of the parallelism of the psychic and the physical as a manifestation of the basic unity of being. Consciousness is conceptual integration, that is to say, integration of percepts up to the highest conceptual achievements. The anatomical substrata of consciousness are the highest centres (frontal lobes), since they represent over again—but in more numerous combinations, in greater complexity, specialty and multiplications of associations—the very same impressions and movements which the lower, and

---

[51] Jackson, *Selected Writings* (note 48), vol. 1, p. 73 (1873).

[52] Ibid., vol. 1, p. 124 (1875).

through them the lowest centres, represent. Although the highest centres represent the whole organism, consciousness is not a function of those highest centres; rather, consciousness is concomitant with their functioning, since the whole brain is responsible.

For Jackson, then, the material basis of the 'organ of mind' is composed of nervous processes representing movements and impressions, and recollection is a 'revivification of such processes which, in the past, have become part of the organism itself'.[53] The key to Jackson's conception of function, therefore, is his theory of representation, which is based on the symbol of a piano accordion.[54] According to this symbol, the bellows are analogous to the cerebellar structures which represent low-level (tonic) movements—e.g., the pump action of the heart or the 'concertina action' of the thorax, whereas the chord buttons and melody keys are analogous to the cerebral structures of the Rolandic (motor) region and the frontal lobes which represent, respectively, the higher-level involuntary (clonic) and voluntary co-ordinations in time and space.

Lowest-level movements are encoded as a continuous tactus or beat (our 'Time Constant'); middle-level movements are encoded as simultaneities or chords; and highest-level movements as successions or melodies.[55] Although the higher centres in the cerebral cortex differ from the lower cerebellar centres as parts representing more intricate co-ordinations of impressions and movements in time and space, there still are two kinds of higher co-ordinations, since there is 'a Melody of Movement as well as a Harmony of Movement'.[56]

---

[53] Ibid., vol. 1, p. 26 (1870).

[54] Jackson's theory is thus a modification of Herbert Spencer's (see above note 48), since he utilised a different model. The scientist and poet, W. S. McCulloch, 'Modes of functional organization of the cerebral cortex', *Embodiments of Mind* (Cambridge, Massachusetts, 1965), pp. 67–71, provided the first hint that Jackson utilised a musical-instrument model, although he did not 'unpack' the model.

[55] See Jackson, *Selected Writings* (note 50), for the Time-constant, e.g., vol. 2, pp. 111–116; for chords, e.g., vol. 1, p. 113: 'The nervous centres represent movements, not muscles; chords, not notes'; for melodies, e.g., vol. 1, p. 261: 'Just as there is an increasing harmony of movement from the muscles to the highest nerve centres, so, no doubt, there is an increasing melody'.

[56] Ibid., vol. 1, p. 261.

Although Jackson continually stressed that nervous processes represent movements, not muscles, he allowed that each muscle may be considered as a single 'note', since muscles in coordination will produce a large repertory of movements encoded in the species and in the individual as many different musical scores.[57]

The second metaphor—that of the tree—was utilised by the neurologist and physician, Wilder Penfield, who conceived the brain stem as an ancient trunk from which, over the centuries, has grown various branches (projections) and, latterly, a leaf-like covering (the cortex).[58] On this conception,

> Each functional subdivision of the cerebral cortex of man may be looked upon as an outgrowth or projection outward of some area of gray matter in the older brain stem ... Thus, the projected area in the newly formed cortex presumably serves to amplify and enlarge a function already being served in some sort of rudimentary manner by the old brain of more elementary animals. ... This [conception of the brain] is in many ways a surer guide by which to predict functional subdivision than the cyto-architectonic parcellation of cortex.[59]

Penfield denied that the cerebral cortex is the site of 'conscious planning'. Rather—and this is his principal hypothesis—there is within the higher brain stem a so-called 'centrencephalic system' that includes all areas of subcortical grey matter, together with their connecting tracts, that serve the purposes of inter-hemispheral, as well as intra-hemispheral integration. This central integrating system is the seat of mind, for it contains 'the record' of the whole 'acquired'

---

[57] For a Jacksonian meditation on the brain's record of experience (is it a script or a musical score?), see O. Sacks, *The Man who Mistook his Wife for a Hat* (London, 1986), pp. 138–142. For a brief contemporary critique of Jackson's theory, see J. C. Kassler, 'Edmund Gurney's *The Power of Sound* (1880): II. Reception', *Musicology Australia* (1995), 18: 13–24, p. 21.

[58] W. Penfield and L. Roberts, *Speech and Brain-Mechanisms* (Princeton, 1959), and W. Penfield and P. Perot, 'The brain's record of auditory and visual experience', *Brain* (1963), 86: 596–693.

[59] Penfield and Roberts, *Speech and Brain-Mechanisms* (note 58), p. 23.

(psychical) stream of consciousness.[60] The 'inborn' sensory-motor systems cannot serve this purpose, because such systems 'make possible sensations and actions which are forever new and changing', so that they 'probably retain, in themselves, little or nothing from the past'.[61]

What, then, is the functional role of the cerebral cortex? The answer is that it works in harmony with the centrencephalic system. For example, during normal brain activity, the 'corticofugal' impulses relating to hearing and seeing pass from the temporal lobes to that part of the neurone record of the stream of consciousness in which hearing and seeing are the prominent components.[62] Hence, the temporal cortex makes possible

> *(a)* the recall of previous perceptions (chiefly auditory and visual) from the past, *(b)* the comparison of past experience with similar present experiences, and *(c)* the subconscious elaboration of signals that interpret present experience for an individual by means of such comparison ... [and, conjecturally] *(d)* ... [the assistance] in the process of adding to, or altering a previously recorded concept, e.g., of ... a person, a voice, or a piece of music.[63]

The brain, therefore, has several neuronal mechanisms that subserve memory—experiential and conceptual. But memory acquisition is by a process of inductive assembly.

For example, a child hears a tune for the first time as a percept—a neuronal 'unit' or 'pattern'. This experience does not enable the child to recognise the tune when heard again. For this to happen, the child must hear the tune frequently, so that he can classify and compare his present experience of the tune with previous

---

[60] This hypothesis is a return to much earlier notions of a common sensorium located in the medulla oblongata; see F. M. R. Walshe, 'Some reflections upon the opening phase of the physiology of the cerebral cortex, 1850–1900', *The History and Philosophy of Knowledge of the Brain and its Function*...ed. by F. N. L. Poynter (Oxford, 1958), pp. 223–234.

[61] Penfield and Perot, 'The brain's record' (note 58), p. 691.

[62] Penfield and Roberts, *Speech and Brain-Mechanisms* (note 58), p. 50, conjectured site in the hippocampus.

[63] Penfield and Perot, 'The brain's record' (note 58), pp. 690–691.

experiences. Little by little, this comparative, classificatory process enables the child 'to learn' the tune—that is, to recognise the tune when heard again. When this happens, the child has formed a concept of the tune by a process of generalisation from particulars. And as the child grows, further abstractions are made, whereby, for example, the tune is classed as of this or that type, this or that structure and so on.

CONCLUSION

From the foregoing, we might conclude (as I hope readers will) that there is a problem for musicologists, one that has historical, theoretical and critical dimensions. I have called this problem 'musical ear'—a shorthand term (like 'green thumb' in horticulture) for creativity and competence in music. The nativist solution defines musical ear as innate concepts or as innate principles or as innate abilities. Hence, nativists lack agreement about the particular content of domain-specific knowledge. Nevertheless, they agree that people endowed with musical ear manifest this gift in infancy. We might ask, therefore: what is the criterion or test of evidence for the innateness of musical ear?

You may recall that Révész answered this question with reference to naming. But a different criterion was given by Galton, who chose reputation as the test for all natural abilities, not just musical ones. And more recently, the cognitive psychologist, Howard Gardner, in seeking to provide evidence for a number of core 'multiple' intelligences, argued for the independence of a music faculty from other faculties on the basis of the 'musical ability' (performance) of precocious children, idiot savants and patients with brain damage following stroke or accident where speech is impaired but not singing.[64] Since all these claims encounter problems of

[64] H. Gardner, *Frames of Mind: The Theory of Multiple Intelligences* (New York, 1983). By stressing the independence of core intelligences, Gardner's hypothesis rehabilitates some of the findings of L. L. Thurston and others, who, in the 1930s, utilised factor analysis (as Galton had done previously). See E. G. Boring, *A History of Experimental Psychology* (2d edn., New York, 1950), p. 576, who summarises these findings as follows: 'there are a great many factors which enter into all sorts of combinations in specific skills, but a few of these

interpretation and empirical support, they constitute one aspect of nativism that deserves historical study and criticism.

But there is another aspect requiring the same kind of study, for nativists admit that the innate gift is undeveloped in infancy. But why characterise infants as having innate musical knowledge? Why not characterise them as having the capacity to acquire concepts, principles or abilities from the domain of music? Or why not argue that the musical sensitivity that an infant manifests may be part and parcel of a more wide-ranging cognitive capacity to organise the world in terms of general concepts, concepts that require mastery of principles and abilities?[65] That these questions raise difficulties for nativists was recognised by Galton,[66] who, in his later study of a 'human faculty', continued to assert that 'the bulk of the respective provinces of nature and nurture are totally different', even though he admitted that 'the frontier between them may be uncertain'.[67]

Allowing that the 'delicate powers' of sense discrimination, though innate, may be developed by practice and education (that the discriminating powers are highest amongst the intellectually ablest, higher in men than women, and higher in Europeans than in 'savages' or 'idiots'), Galton still believed that the 'long sequences of changes in character ... are regulated by internal mechanism' 'like the tunes of a musical snuff box'; that is to say, those changes are due mainly to 'nature', even though 'nurture' may act before birth, 'causing the potential faculties at the time of birth to be in some degree the effect of nurture'.

---

factors are more generally involved than others in successful human activity and these may be regarded as primary mental abilities'. Fodor's modularity thesis is different, since it stresses the isolation and domain-specificity of processing; see J. Fodor, *The Modularity of Mind* (Cambridge, Massachusetts, 1983).

[65] In the case of our child composer, X, his grown-up interests in artificial intelligence suggest that music was the sign of other more general abilities in, e.g., pattern recognition and applied logic. For example, a tape recording, made at the age of four, suggests that X possessed procedural knowledge (know-how) relating to pattern recognition: he played various five-finger sequences of notes on the piano and claimed ownership of the sequences—they were 'my melodies'.

[66] F. Galton, *Inquiries into Human Faculty and its Development* (London, 1883), pp. 177–178.

[67] Ibid., pp. 33, 37, 181–182.

With these statements Galton seems to have capitulated to the empiricists, who do not deny that humans come equipped with inborn capacities but assume that those capacities are instruments or mechanisms for the acquisition of knowledge from sense perception by inductive assembly. For them, musical ear is defined as an accustomed or practised 'ear' that is determined psychologically or physically or biologically. But this solution has its own set of problems that deserve historical study and criticism. For example, empiricists agree that what is brought to the task of acquisition is inductive assembly. But how does the associative compiler know what to assemble together? Is the accumulation of knowledge as slow and uncertain as inductive assembly implies? And how does the assembling compiler know what the 'game' is about (e.g., naming)?

Finally, if traditional solutions to the problem of musical ear are unsatisfactory, what, then, constitutes a satisfactory solution? Here, then, is a challenge not only for philosophers and psychologists but also for musicologists.

INDEX

For Product Safety Concerns and Information please contact our EU representative GPSR@taylorandfrancis.com
Taylor & Francis Verlag GmbH, Kaufingerstraße 24, 80331 München, Germany

www.ingramcontent.com/pod-product-compliance
Lightning Source LLC
LaVergne TN
LVHW010621300926
849162LV00001B/22

* 9 7 8 0 8 6 0 7 8 8 6 2 1 *